Meanderings in the Bush

Natúral History Explorations in Outback Australia

February 2011
Patrick:
Happy Meanderings
Throughout retirement –
Dick and Barbara

Richard E. MacMillen
with Barbara J. MacMillen

Front cover photo: More than 300 Little Corellas (cockatoos) resting in a dead gum tree on the bank of the Diamantina River, near Birdsville, central Australia, 1992.

Back cover photo: Barb and Dick MacMillen, at Government Hole (a billabong) on Mabel Creek Station, near Coober Pedy, South Australia, 1989. Photo by Ian MacMillen, age 5.

First published by Dog Ear Publishing
4010 W. 86th Street, Ste H
Indianapolis, IN 46268
www.dogearpublishing.net

ISBN: 978-159858-383-0

This book is printed on acid-free paper.

Printed in the United States of America

This book is dedicated to the lasting memory of Russ, whose meanderings were cut far too short by years

Russell V. Baudinette, 1945-2004

Proceeds from sale of this book will help support The Russell Baudinette Travel Scholarship for Ph.D. students of Animal Physiology and Ecology at Adelaide University

Figure 1. Russ Baudinette and Dick MacMillen, Adelaide, South Australia, 2002.

TABLE OF CONTENTS

ACKNOWLEDGMENTS

Many people, agencies, and institutions have aided in the foundations for and preparations of this book. First and foremost is my wife and coauthor of Chapter 4, Barb, who stood beside me throughout the most important episodes that culminate herein. In addition to being mother, wife, and friend, she participated in many phases of the scholarship, withstood difficult conditions, nursed me through asthma and severe burns, propped me up during six years of a demanding administrative assignment, and brought us our son Ian. In turn, Ian helped to rekindle the flames of adventure and discovery with his boundless enthusiasm, energy, and curiosity. Except where noted otherwise in the captions, all photographs were taken by Dick or Barb MacMillen; the maps were rendered electronically by Barb MacMillen, using downloaded shapefiles from Geoscience Australia.

I thank my several Australian friends and colleagues, all fine scholars, who, collectively, read and commented freely on earlier portions of the manuscript, improving their contents and composition; these are Russ Baudinette, Terry Dawson, Lynn Day, Margaret Friedel, Peter Greenaway, Dick Kimber, Tony Lee, Steve Morton, Alan Newsome, and Pat Woolley. Here in the U.S. the manuscript benefited greatly from the careful reading and comments of Barb MacMillen, Jerry Burke, Dave Hinds, and Julie Monson. For errors, oversights, and exaggerations only I am to blame.

Invaluable aid, advice, and friendship were provided by the good people of central Australian livestock stations: Reg and Marg Arthur, Norman (PeeWee) and Diane Clarke, and Geoff and Shirley Schrader of Sandringham and Glengyle Stations; Bill Stalley of Rossmore Station, and Shane O'Connell of Mabel Creek Station. Thank you for sharing your lands, your wisdom, and your welcome.

Financial aid and/or logistical support came from a variety of sources and in varied form over the years, and include: the Fulbright Foundation, the National Science Foundation, the Australian Research Council, the Flinders University Board of Research, the University of California Education Abroad Faculty Research Exchange Program, the Mark Mitchell

Foundation, the American Philosophical Society, the National Geographic Society, Pomona College, the University of California, Irvine, Monash University, the University of New South Wales, the Flinders University of South Australia, Harry Recher and the Australian Museum, the CSIRO Centre for Arid Zone Research, and S. Kidman & Co. To all of these I am forever grateful.

Permission to collect specimens and, in some instances, the provision of specimens was granted by the following Australian agencies: National Parks and Wildlife of New South Wales, National Parks and Wildlife of South Australia, Queensland National Parks and Wildlife, and the Conservation Commission of the Northern Territory; I thank them for unfailing cooperation.

And finally, an explanation and a confession that, I hope, will also stand as an acknowledgment. Those who write and read papers in scientific journals might agree that the style of presentation is very stereotyped, and not particularly lively, and that attempts at enlivenment are usually inhibited by editorial scrutiny. These constraints on expression clashed with some of my maverick ways, and I found relief in the Acknowledgment sections of my journal articles, including those based on Australian work. It is no secret that I like beer, especially Australian beer, so in one way or another I have managed to sneak in a plug for a brew that contributed to the successful undertaking of a piece of work. Hence, expressions of thanks to The Coopers Trust (Coopers Real Ale of South Australia), T.O.O. Heys (Toohey's Lager of New South Wales), Stubby Southwark (= a short bottle of Southwark Bitter of South Australia), and G.I. Nebra (for ginebra which is gin in Spanish, a brief excursion into the harder stuff!). Thank you editors for letting me have a little literary fun!

INTRODUCTION

I love a sunburnt country,
A land of sweeping plains,
Of ragged mountain ranges,
Of droughts and flooding rains;
I love her far horizons
I love her jewel-sea,
Her beauty and her terror -
The wide brown land for me!
Dorothea Mackellar 1911

These simple lines of a poem that is learned by nearly every school child in Australia is prophetic of my fascination with that country. Prophetic, because I only learned of the poem long after my fascination had developed. The fascination was first inspired by my postgraduate mentor George Bartholomew, from the University of California, Los Angeles, who spent sabbatical leaves in the 1950s and 1960s in Perth and Brisbane, and who, by his own work there, showed it to be an unusually fertile arena for students of physiological ecology. Many of his former graduate and postdoctoral students, and their academic progeny developed this fascination nearly to the point of an obsession to make at least one sojourn to this sunburned country. I have probably been as guilty as anyone in developing this obsession to a near addiction, and feel that an explanation is in order by way of this volume.

My personal fascination with Australia is that, as Dorothea Mackellar implies above, it is an arid land, and one of great climatic unpredictability. In addition, as an island continent in the southern hemisphere that owes its biogeographic origins to Gondwana and subsequent continental drift, it has a unique and diverse biota, wholly different from northern hemispheric images. Gondwana was a gigantic, southern hemispheric "supercontinent" made up of the land masses that we know today as South America, Africa,

India, Antarctica, and Australia; it began to break up in the Early Cretaceous, about 135 million years before present, and the resulting "new" continents began slowly to drift to their present locations, leaving Australia isolated from the rest as an island continent (Heatwole 1987). Therefore it is an unusual biological arena for studies of the impacts of resource limitation on native floras and faunas that have developed in isolation from those of the northern hemisphere, upon which much of the prevailing ecological wisdom is based.

For the physiological ecologist the Australian arid zone has been particularly rewarding, for much of our focus has been on the roles of energy and water in the lives of organisms. Prevailing aridity, by definition, results in a paucity of water that, concomitantly, yields a reduction in primary production, i.e. energy (Rosenzweig 1968). Thus, both energy and water may be severely limiting simultaneously. Physiological ecologists, who have specialized in studies of land-based organisms attempt, through a combination of laboratory and field studies, to explain how organisms that live in seemingly inhospitable places, thrive there. To many humans, central Australia appears to be just such an inhospitable place, largely because of its reputed aridity. However, when the infrequent rainfall does occur in central Australia, it may be torrential, resulting in widespread flooding, and inundation of vast areas of otherwise desert landscape (Allan 1990). Such rainfall and flooding results in a temporary respite from the prevailing aridity, with ensuing but short-lived flushes of primary and secondary production. Thus, it is a land of extremes of water and energy availability, and therefore an ideal natural laboratory for those of us interested in the relationships between these commodities and the organisms that rely upon them.

Consequently, I have returned repeatedly to Australia to concentrate on studies of energy and water regulation in its animals. There I have been affiliated with a variety of institutions and colleagues, all of whom have been instrumental in instructing me in the ways of life in the bush. My initial introduction to Australia was with Tony Lee, John Nelson and Russ Baudinette at Monash University, near Melbourne in Victoria during 1966-67. This was followed in 1974-75 at the University of New South Wales, near Sydney, with Terry Dawson. In 1983, a year was spent affiliated with Harry Recher and the Australian Museum in Sydney, but with nearly all of the time devoted to work on Dangar Island, on the Hawkesbury River north of Sydney. There, I was accompanied by my wife Barbara, her sons Chris and Steve, and it was there that our son, Ian, was born. In 1985 Barbara, Ian and I spent five short weeks with Russ Baudinette at Flinders University of South Australia, and then returned to Flinders for additional work with Russ for six months in 1989. The initial drafts of this book were undertaken in 1992 when Barbara, Ian and I spent three months in Alice Springs, Northern Territory, where I was affiliated with Steve Morton and the Commonwealth Scientific and Industrial Research Organization's (CSIRO) Division of Wildlife and Ecology, and its Centre for Arid Zone Research. Although nearly all of my Australian research has focused on arid-zone physiological ecology, I have been attached mainly to institutions located in moderate, coastal environments, since that is where the bulk of the nation's population and academic institutions are located. It is both fortunate and fitting that we are able to initiate this synthesis of our geographic and scientific meanderings through the Australian bush from the vantage point of Alice Springs, central Australia.

In addition to attempting to reflect something of my fascination with Australian natural history, I hope that we are able to convey information that seldom appears in print in the scientific literature. This information constitutes the logistical and personal aspects of experimental specimen location, observation, and collection, including the exploration and human interactions that are involved in this process. In my line of research, less than fifty percent of the time and energy involved is in data acquisition and reduction, and writing, yet that is the only part that appears in the published manuscript. The major part of the effort, which remains pretty much unmentioned, is devoted to exploring, locating, and collecting experimental samples of wild animal populations, convincing authorities and landowners that you should be allowed to collect them, dealing with the countless unanticipated events that invariably occur (a lot preventable, some completely accidental), and just plain having fun. Add to this unfamiliar topography, climatic unpredictability and human vagaries, and adventure will most likely proceed. It is this adventure and the resulting scientific observations and conclusions upon which we will focus herein as we meander through the Aussie bush.

I thought that a clarification of the title is probably in order. My Webster's dictionary (Friend and Guralnik 1960) defines the verb meander in the following way: 1) to take a winding or tortuous course; 2) to wander aimlessly or idly; ramble. My intent throughout (this book and in the better part of my life) has been to follow the former definition, but I will be the first to admit (and pleasurably) that there are occasional forays into the latter.

The bush is a more difficult concept for me to define for it is a British colonial expression I have heard in Australia, New Zealand, and South Africa. The nearest American counterpart is "boonies" or "boondocks" which in Webster's is: 1) a jungle or a wild, heavily wooded area; wilderness; 2) any remote rural or provincial region; hinterland. The Aussie term

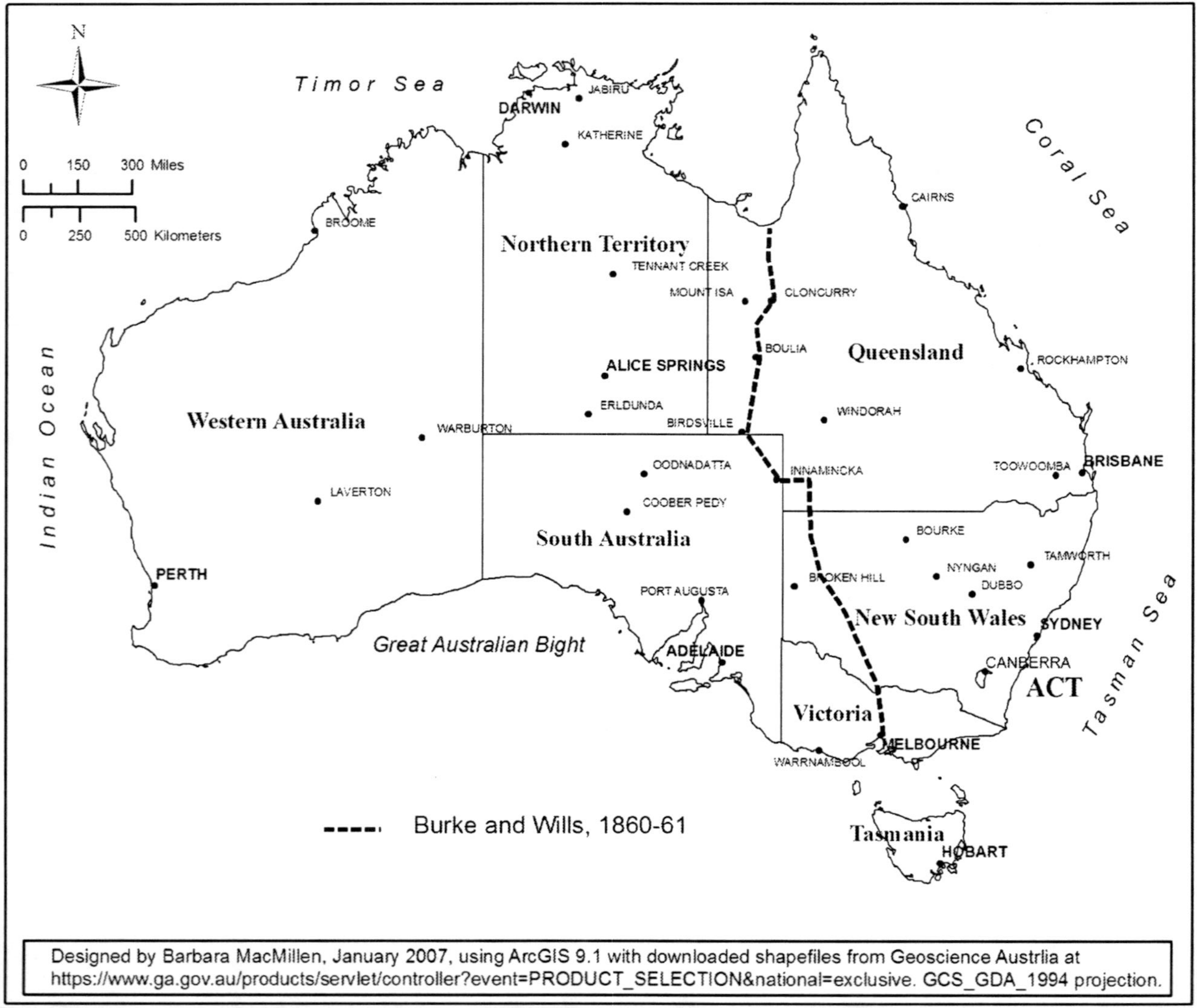

Map 1. Australia, its states, territories, cities, and towns. Also included is the approximate route of the ill-fated Burke and Wills exploratory expedition that commenced in Melbourne, 1860.

"bush" incorporates all of these meanings, too, depending upon where you are from and where you are going. For a Sydney (population 3,460,000) resident, "going bush" may simply be going to the next smaller city in New South Wales, such as Newcastle (population 264,500). A Newcastle resident might be going bush on a visit to Orange, N.S.W. (population 33,200), while Orange residents would speak of Nyngan, N.S.W. (population 2,500) and Bourke, N.S.W. (population 3,200) as going bush. Much of the bush through which this book meanders is "Back O' Bourke" including tiny outback villages such as Birdsville (population 30), Windorah and Bedourie in Queensland, and remote desert wilderness between them and beyond. It is interesting and perhaps heartening from a naturalist's perspective that Windorah and Bedourie are not even blessed with entries in the 1991/92 Summer edition of The Royal Automobile Club of Victoria's "Accommodation Australia" guide, the source of the foregoing population figures.

It was Frederic Crescitelli, a former physiology professor at UCLA, who first introduced me, as a graduate student, to the term serendipity, which he described as "fortuitous good luck". (For a scholarly history of the word, serendipity see Morton and Barber 2004). Dr. Crescitelli went on to explain how many of the great medical and scientific breakthroughs have been serendipitous, such as the discovery of penicillin and of the utility of Teflon coating, to name a couple. He attributed at least a portion of his own professional career as a cell biologist to serendipity, and it has become apparent to me that it also plays a very important role in the science of ecology. For an ecologist, in particular, serendipity means being at the right place at the right time, and being aware of it. This requires reasonable powers of perception and deduction that are two of the prerequisites to success in science. But ecologists are blessed (or cursed) with the vagaries of nature, which fluctuate in seemingly peculiar ways, yielding every now and then wonderful opportunities for ecological investigation to the lucky and perceptive individual who happens along.

Certainly, my Australian work has benefited greatly from serendipity, especially in the earlier phases when I was less familiar with the fauna, its natural history and distribution. My initial plans while preparing for my first visit to Australia (1966-'67) were to continue my earlier studies of water and energy regulation in desert doves, by conducting experimental work on the Spinifex Pigeon (*Petrophassa plumifera*), a handsome arid-zone bird very similar in habits to North America's desert-dwelling Gambel's Quail. Shortly after my arrival, I had a chance meeting with Basil Marlow, then Curator of Mammals at the Australian Museum in Sydney, who informed me of large populations of the eutherian hopping-mouse *Notomys* spp. (Family Muridae) in southwestern Queensland, on the edge of the Simpson Desert. Basil had recently returned from this arid region and was quite excited about finding the rodents that resemble, but are not even closely related to, North American kangaroo rats (*Dipodomys* spp), on which I had worked earlier as part of my doctoral dissertation. He thought the *Notomys* merited study, and they certainly sounded attractive to me as an alternative possibility. As it turned out, although I found a large population of Spinifex Pigeons near Mount Isa, in western Queensland, they were nearly impossible to capture alive in sufficient numbers. The birds were widely dispersed due to recent rains, abundant, widespread surface water, and food; they simply could not be lured into traps. The fortuitous alternative possibility of studying *Notomys* became a reality, and resulted in three often-cited publications (MacMillen and Lee 1967, 1969 and 1970), whose conclusions are summarized in Chapter 1.

Chance encounters after rains (which seem to follow me everywhere in Australia, including the Centre) with the curious Inland Crab, *Austrothelphusa (Holthuisana) transversa,* eventually led to another serendipitous study, and became a continuing vehicle to eminence of my collaborator Peter Greenaway of the University of New South Wales, as described in Chapter 3. Another completely fortuitous study that created probably the greatest stir of my early contributions to the scientific literature also was conducted at Monash University during my first Australian visit, on the metabolic rate of dasyurid marsupials, with John Nelson. Dasyurids are curious (to those of us from the Northern Hemisphere) insectivorous and/or carnivorous marsupials that range in size from tiny shrew-like critters to carnivores the size of a dog, and they occur across a broad range of habitats. All of them have long snouts and muzzles that bear a healthy set of sharp teeth; their Northern Hemisphere counterparts, although only extremely distantly related, are the insectivorous shrews and the carnivorous weasels, skunks, badgers, and wolverines, all of which are eutherian mammals. John Nelson had a captive colony of the Northern Quoll (*Dasyurus hallucatus*) at Monash, housecat-sized dasyurids from northern Australia. And nearly everywhere we turned dasyurids of other sizes would appear, brought into the University as chance captures or called to our attention as being available on loan as captives, or captured by us while in pursuit of other animals.

I had been aware that information was available on the relationship between metabolic rate and body mass in eutherian mammals (Brody 1945; Kleiber 1932). Eutherian (AKA placental) mammals are those with which I was most familiar, and that, unlike marsupials, have long gestation periods while the developing fetus is plentifully nourished through the maternal placenta. The relationship between metabolic rate and body mass in marsupials, whose young are born in a nearly embryonic state without much intrauterine nourishment, was unknown. As I had brought with me on sabbatical leave equipment necessary for the measurement of oxygen consumption (=metabolic rate), it dawned on me that here was a unique opportunity to collect data on the relationship between metabolic rate and body mass within a single family of marsupial mammals (the Dasyuridae) that varied in size more than 500 fold. Therefore, John Nelson and I took advantage of this array of dasyurids from a variety of sources, including generous loans of precious specimens, and were able to make metabolic measurements of twelve species (MacMillen and Nelson 1969). We found that the resting metabolic rates of our dasyurids were only about two-thirds those of similar-sized eutherians. It was equally fortuitous and completely unknown to us at the time, that Tony Hulbert and Terry Dawson, at the University of New South Wales, were conducting a parallel study on a taxonomically more diverse array of Australian marsupials (Dawson and Hulbert 1970). Fortunately, for all of us, the answers and conclusions of the two independent studies were nearly identical, and, as fortune might dictate, our shared independent interests led to later personal collaborations with both Terry Dawson and Tony Hulbert, and to lasting friendships.

In addition to the geographic and scientific meanderings that are reflected directly in this book, I see, upon rereading the several drafts, that it also reflects in a less direct manner the many personal and philosophical meanderings that have shaped my life over the last third of a century. In 1966 I arrived in Australia well-versed in the classical physiological ecology of the time, in which one looked for bizarre species and attempted to unravel their special attributes, one species at a time, without addressing the community of organisms with which they coexisted and were frequently interdependent. In addition, I arrived well versed in North American natural history, but with only a travel magazine knowledge of Australian natural history. Since those naive days, and particularly with the advent of community ecology, I have come more and more to view an organismal population as one of many elements of an interwoven community of organisms, each of which impinges in some manner upon the others. My continuing exposures to the Australian bush, together with extensive reading, have greatly expanded my grasp of its natural history nearly to the point of intimate familiarity and belonging. I hope that these transformations have added a degree of sophistication and insight to my later Australian work. A related transformation that is indirectly reflected through the earlier portions of the book, and that culminates in Chapter 8, is a growing awareness of the extent to which human activity changes the landscape, too frequently to the detriment of nature. The truths of community ecology - that the organisms of a community interact and are frequently interdependent - helped me to shape an appreciation for biodiversity. My laboratory and field work revealed to me that animals (and plants, too) that are robust members of their pristine communities may be weakened to the point of extinction when those communities are perturbed by human activity, thereby reducing species richness and community integrity. These awarenesses have left me with the uncomfortable conclusion that humans are disassembling in a couple of centuries much of which nature fabricated over the last billion years.

A final introductory comment has to do with a distinct shift in attitudes I have experienced and perceived during my career - that towards greater professional equality between the sexes in ecology and natural history, and from which I have benefited greatly, particularly in recent years. On my initial visit to Australia in 1966-67, I was a guest of Monash University, where A.J. (Jock) Marshall was Professor of Zoology, and my official host. Jock Marshall and Russell Drysdale had recently published their famous book of an Australian biological expedition, "Journey Among Men" (Marshall and Drysdale 1966), which I read avidly at the time and without really considering whether the expedition was affected by the lack of female participation. I still remember wondering at the time about their dedication "For Bon and Janey who were kind enough to stay at home." I did not get to know Jock well during my stay at Monash for he was nearing the wrong end of a bout with cancer, so I am not familiar with his gender views. Nevertheless, I was aware that both Australia and the United States had a long way to go in achieving gender equality in virtually all endeavors, including academics. Progress has been made, including my own transformation, but there is still a long way to go. My primary reason for mentioning this, in what I hope will not be construed as mere tokenism, is that nearly all of my Australian field work has benefited professionally from participation by both sexes, for which I am grateful. Especially gratifying has been the success of the work undertaken in Australia beginning in 1983, with the fullest collaborative involvement of Barbara MacMillen, who even brought us our own bush baby, Ian.

SUGGESTED FURTHER READING:

Dawson, T.J. and A.J. Hulbert. 1970. Standard Metabolism, Body Temperature and Surface Area of Australian Marsupials. American Journal of Physiology 218: 1233-1238.

MacMillen, R.E. and J.E. Nelson. 1969. Bioenergetics and Body Size in Dasyurid Marsupials. American Journal of Physiology 217: 1246-1251.

Chapter I

"BOGGER DICK" AND THE ROAD TO SANDRINGHAM, 1966

When you come from the northern plains
Where the girls and grass are scanty
Where the creeks run dry or ten foot high
And there's either drought or plenty.
Pass the billy around boys
Don't you let the pint pot stand there
For tonight we'll drink the health
Of every overlander.
Traditional Bush Ballad, ca. 1850

Stimulated in part by my conversation with Basil Marlow about large populations of native hopping-mice (Rodentia: *Notomys*) in southwestern Queensland, and, in part, by learning that the Spinifex Pigeon also occurred in that sector of the country, I proposed to Tony Lee that we mount a collecting expedition to obtain experimental animals. By returning with both or either species to Tony's lab at Monash University, we could then proceed with our collaborative studies of the physiological attributes that contribute to their successful arid-zone existences. At the time Tony was a Lecturer in the Department of Zoology and Comparative Physiology at Monash, near Melbourne. He and I had been graduate students together at UCLA, where we earned our Ph.D.s under the guidance of George Bartholomew. Tony is Australian, with a Bachelor's Degree (with Honours) from Melbourne University, and subsequent graduate work at the University of Western Australia prior to his arrival at UCLA. At that time (1966), we were both brash, young (mid-thirties) neophyte academicians; now we're older and retired!

Tony was enthusiastic about the prospect of a collecting expedition to the Queensland Bush and arranged for use of a four-wheel drive, long-wheel base Toyota Landcruiser that belonged to his Department. He proposed further that we be accompanied by Mike Waldman, a Departmental Ph.D. student in paleontology from Great Britain, and Dione Gilmour, a close friend of Tony's, the wife of another Departmental graduate student and an employee of the Australian Broadcasting Commission; neither Mike nor Dione had ever gone bush before, nor had I (at least in Australia). Mike was interested in dinosaurs, and there was the chance that our trip could traverse some known fossil fields; for Dione the trip would provide an opportunity to develop ideas for future nature programs, an endeavor for which subsequently she has become renowned.

It was August, late winter, and theoretically an ideal time for trekking in the outback, with typically cool temperatures and lowest rainfall (Kalma and McAlpine, 1979). Our destination was to be two cattle stations north of Birdsville in southwestern Queensland that I had learned about from Basil Marlow, Sandringham Station and Marion Downs. I had written to the managers of each, indicating our probable arrivals. In retrospect I think I was inwardly more excited about the possibility of locating hopping-mice for study than Spinifex Pigeons, as I had always been attracted more to furry than to feathered critters. And the thought of arid-adapted, seed-eating rodents living in Australian deserts otherwise dominated by marsupials really intrigued me.

After several days of preparation and packing supplies we were off, the four of us and our gear barely fitting into the Landcruiser. Since we were heading for the arid Centre, and it seldom rains in August, we did not include a tent, and would sleep out under the stars. Luckily, we had the foresight to include a large tarpaulin!

Our route was to take us north from Melbourne through the state of Victoria, continuing north through New South Wales to Bourke, all on fairly well established, largely sealed (paved) roads and highways. Bourke is traditionally the jumping off point for the outback, or "Back O' Bourke". From there we would head north into Queensland, then northwest along a network of dirt tracks through southwestern Queensland to Birdsville, and then north to Sandringham, Marion Downs and then Mount Isa, before heading south again to Melbourne (see map 2).

Without my knowing it the route we had chosen was to converge on that of the ill-fated Burke and Wills expedition, which ended with their deaths on Cooper Creek a little over 100 years earlier (see Anon. 1861; Moorehead 1985). Their exploring expedition, which was to be the first south-north crossing of the continent by Europeans (see map 1), left Melbourne with a crew of men, Camels and horses on 20 August 1860. It arrived at Cooper Creek in central Australia on 11 November 1860, where a permanent supply depot was established just east of what is now Innamincka, on the South Australia-Queensland border. On 16 December, in midsummer, leader Robert O'Hara Burke, astronomer-surveyor John

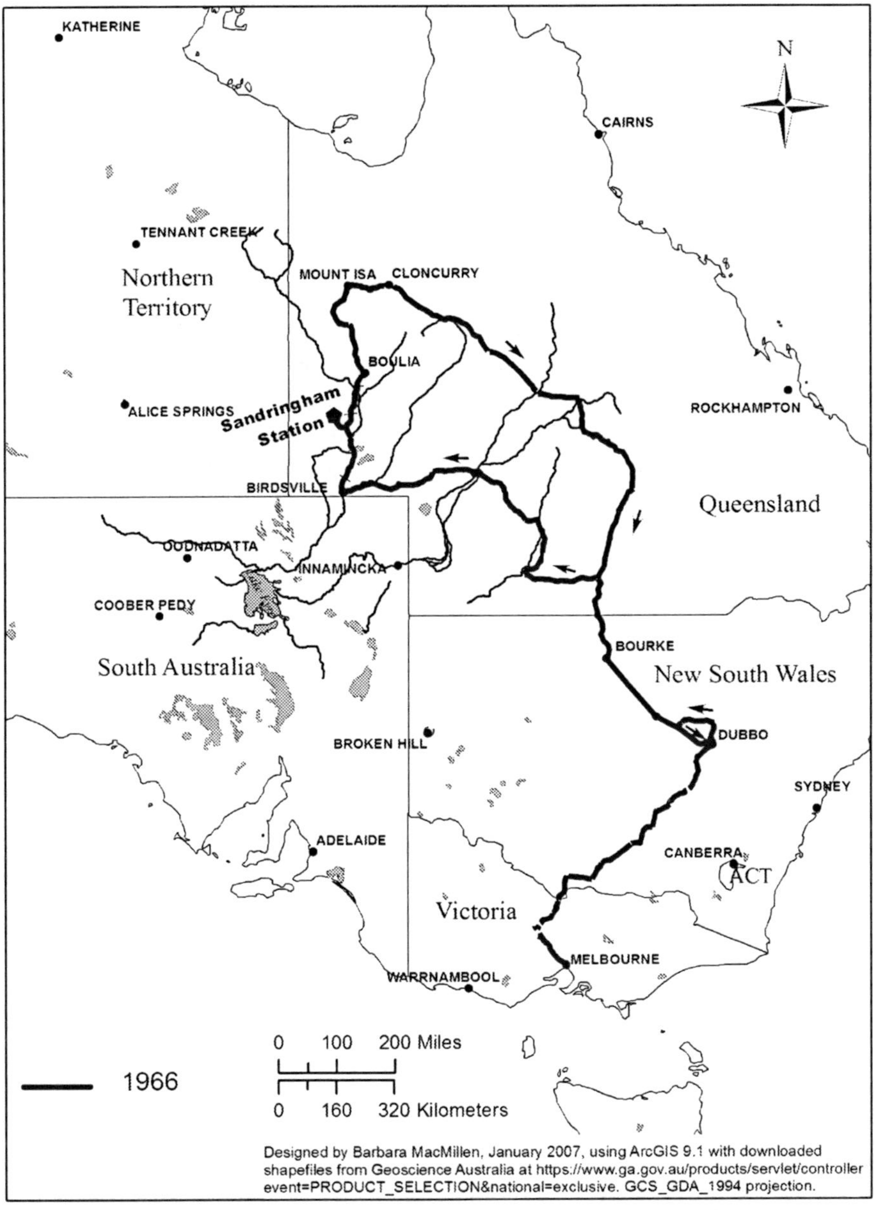

Map 2. The route of our 1966 collecting expedition from Melbourne to Sandringham Station and return.

William Wills, with Charles Gray and John King departed the supply depot for the north and the Gulf of Carpentaria, across uncharted desert, leaving behind at the depot most of the men and supplies with orders to await their return. Their forced march to the Gulf left them malnourished and exhausted when they reached it on 10 February 1861. On their return to the depot Gray died, several animals died or were abandoned, and Burke, Wills, and King arrived back at the depot on 21 April, only to find that it had been abandoned by the support party on the same day, 8-10 hours earlier. Too weak and disheartened to follow, they attempted to eke out an existence on the meager supplies left behind and whatever they could glean from nature. First Wills and then Burke died on Cooper Creek, leaving King alone, whose survival was owed to eventual adoption by local Aboriginals. King was finally rescued by a search party on 15 September 1861, and returned to Melbourne a weak, confused hero on 23 November 1861, the only surviving member of this first group to cross the continent. Our path from Birdsville north to Mount Isa was to parallel closely part of the explorer's route between the supply depot and the Gulf of Carpentaria (see map 1).

The first night of our collecting expedition was spent near Echuca in northern Victoria, and, in spite of it being the "dry season", it rained nearly all night. Luckily for us, we had the tarp, which was nearly waterproof and, when pulled over all of our sleeping bags, kept us fairly dry. In response to my query to Tony that I didn't think it was supposed to rain in August, he replied tersely, "Well, it's not supposed to!" It proceeded to rain on and off for the next two or three days, which made preparing meals and sleeping damp and awkward, but traveling wasn't impaired as long as the roads were sealed.

Our first adventure (or misadventure) of the trip came on the second night when we called in at the Doyle sheep property near Gulargambone, northwest of Nyngan in central New South Wales. This was a station on the Bogan River where Tony had been conducting research on arid-zone frogs, and where he had arranged for us to be quartered out of the rain in the sheep-shearing shed. Upon our arrival we learned that a Doyle son was celebrating his 21st birthday, calling for a country party and dance at the homestead, to which we were enthusiastically invited. Twenty-firsts are not taken lightly in Australia, and friends and relatives came from miles around to join the festivities, fueled by large amounts of strong Aussie beer. Music for dancing was provided by a record player, and the dancing became increasingly vigorous as the evening proceeded. Probably fortunately for all of us the party came to a sudden halt about midnight when Dione and Mike, while engaging in some particularly energetic footwork, managed to upset a mug of beer that had been set on the record player, spilling its contents into the instrument. This resulted in short-circuiting the entire electrical system in the homestead, casting it into quiet darkness, and people said their good-byes as they drifted off toward their homesteads or to quieter activities under the cover of darkness. After making our apologies, which were accepted with good cheer, we made our way to the shearing shed and into our sleeping bags.

The next morning we headed north, to enter the region "Back O' Bourke", and onto dirt roads and tracks. Optimistically we headed onward knowing full well that we were entering the real outback where it seldom rains, especially in August. We proceeded from Bourke northward to Cunnamulla, Queensland along a damp, but solid, graded dirt road without incident, and turned west toward Thargomindah. This was lovely country with stands of River Red Gums (*Eucalyptus camaldulensis*) along the creek beds, and with the damp, red earth nearly glowing on all sides. Except for the river gums, much of the land between the creek beds was heavily overgrazed and devoid of vegetation, as the present period of rainfall had been preceded by a lengthy period of drought. It was not raining as we proceeded eastward and we were making good time along the dirt track, until we arrived at the Paroo River. The Paroo had a ford crossing, and since it was running a little water, we stopped before attempting the crossing to determine its depth. Tony decided to wade into the ford to explore the feasibility of the crossing and, as he did so, the water in the river began to rise at an alarming rate. He barely made it back to the bank, as the water proceeded to rise, and rise. We were witnessing a genuine flash flood! Before long trees were floating down the river. This was more than enough to persuade us to back off to high ground, and make an early camp for the evening. The Paroo River, which flows south eventually to join the Darling River in New South Wales, has a fairly extensive watershed to the north of where we were, and where it must have rained heavily, resulting in the flash flood we witnessed.

Had we been better informed of the topographical features that lay ahead of us we would have realized that we were about to enter the "Channel Country", comprising the parallel and vast watersheds of Cooper Creek, the Diamantina and Georgina Rivers which flow from the north and northeast to southwest toward Lake Eyre, in South Australia (see map 3).

The Lake Eyre Basin is the largest inland drainage in Australia, covering well over a million square kilometers of surface area, and comprising about one-quarter of the continent (Allan 1990). As such, during periods of heavy rainfall as we were experiencing, runoff water may be transported scores to hundreds of kilometers from the point where it was produced as rain. Our experience at the Paroo, with its smaller drainage, should have served as a warning, as should have also the old wooden boats we were occasionally to see tied to gum trees way up the banks of nearby dry watercourses. These we were told were used by the locals to cross the rivers during flooding, and were always kept on hand because local weather

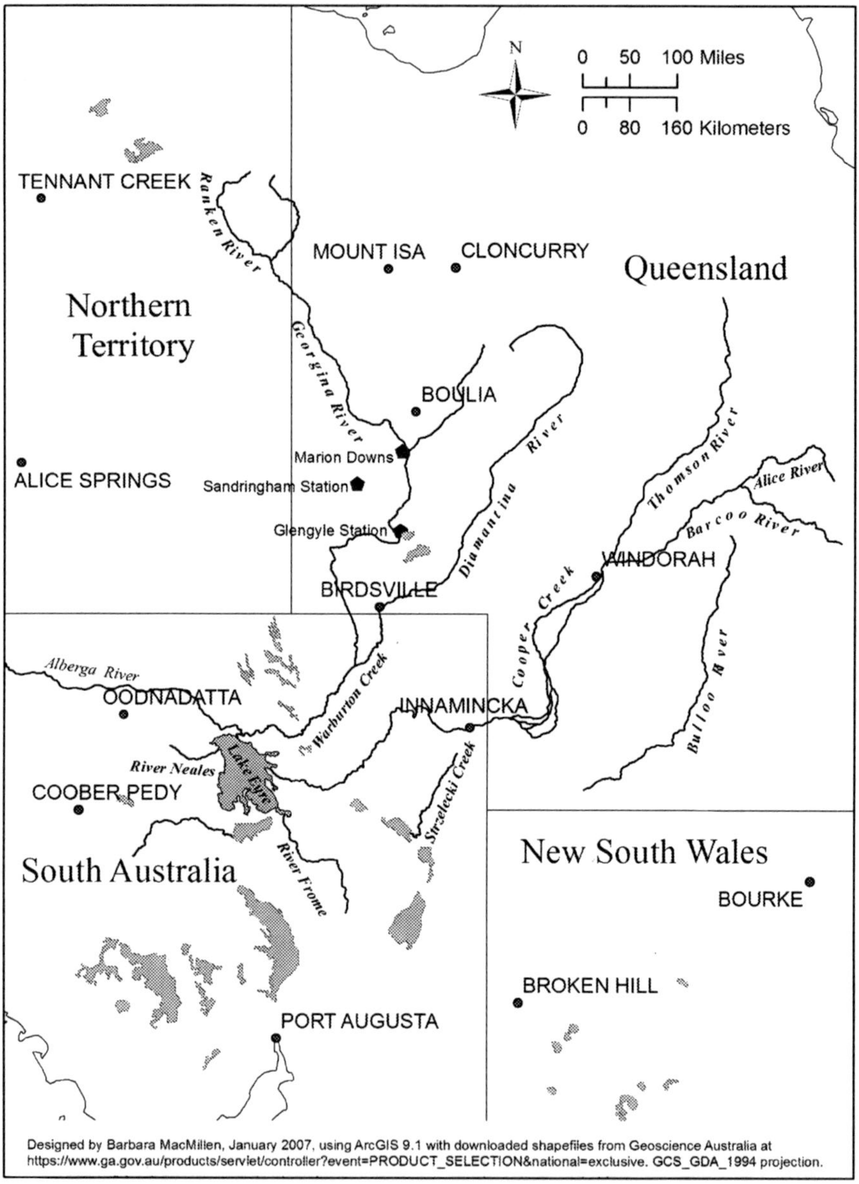

Map 3. The Lake Eyre Basin, the northern one-half of which comprises the Channel Country. During flooding in the Channel Country, the rivers and their tributaries overflow their banks, turning the entire country into a temporary inland sea.

conditions did not necessarily reflect what was happening further up the drainage. I was beginning to understand that Australian deserts were different from those I was used to back home, where occasional flash floods do occur, but they are usually localized.

By morning the river had subsided back to a trickle and we were able to cross without incident to Thargomindah, and then north to Quilpie whence we headed west toward Birdsville, and very remote country. From Quilpie we proceeded to Windorah where we crossed Cooper Creek on a bridge, much to our relief. Windorah looked like a town out of a western movie, with a shop or two and a few wood-frame houses lining the dirt street, and a pub that substituted for a saloon and a gas station. In central Australia, the main centers of commerce and communication are the pubs, formally entitled hotels, but where the sale of beer outnumbers the sale of meals and rooms by a very wide margin. Because of very thirsty conditions and because of their sparse distribution we were forced to stop at nearly every pub we encountered. At each pub, we were put through the traditional ritual of being observed quizzically as strangers, with the unspoken expectation by the locals of our telling them our story. We were surprised when our story of looking for mice and birds to collect was accepted rather matter-of-factly. When we arrived at the Windorah Pub the barman looked up at us and said knowingly, "You're the rat catchers!" The others at the bar looked at us knowingly too, and nodded. We had been accepted as odd members of their society, and our arrival was being announced in advance, likely by two-way radio! While seldom volunteering much information, these bush people were quietly cordial and most helpful when asked. They seldom inquired of us, but seemed genuinely interested when we explained what we were up to.

West of Windorah the dirt track had been badly churned by semi-trucks and trailers that, in combination with rainfall and runoff water, had turned the track into two parallel grooves of mire, especially in the low places. I was elected to drive through what seemed an eternity of this muck. Fortunately the Landcruiser was a powerful brute and, when put into second speed, low-range, four wheel drive would scream through the mire with the tires careening between a myriad of parallel grooves, often obliterated from sight by standing water, until another high point was reached where we could stop for a breather. It was hard, nerve-wrenching work, and the cold beers waiting for us at the Betoota Pub (if it was open!) became more enticing as the hours passed.

At one of our stops west of Windorah, and along a small creek channel with small, standing pools of water, we noted vertical holes in the mud about 3 cm in diameter. Tony Lee, who had previously done some very interesting work on Australian desert frogs, suspected that the holes might belong to these strange critters. These frogs spend the lengthy times of drought burrowed underground, and come out to feed and reproduce only during and after the infrequent rains; conditions comparable to what we were witnessing. Therefore, we got out the shovel and commenced to dig, following one of the holes down into the soft clay mud. Upon depositing one of the shovels-full of ooze on the sandy bank, we saw movement. Upon extricating the source of the movement, we were amazed to see, not a frog, but a small crab, resembling at least superficially, a shore crab. I was prepared to accept the existence of a desert frog, as we have them in North America, but I was wholly unprepared to be confronted by a desert crab! Here in the middle of the Queensland desert, several hundreds of miles from the sea, where self-respecting crabs ought not to live, was a live, seemingly healthy crab living in fresh water. What do they do when the country dries up, which is most of the time? This will not be the last time we will encounter the Inland Crab in our meanderings, and each time my curiosity was tweaked further, inviting investigation. Here, too, was my initial awakening to Australia as a land of ecological wonder and opportunity, where the unexpected comes to be the expected.

As we neared the isolated bush pub of Betoota, situated between Cooper Creek and the Diamantina River, we entered a mile-wide flood plain, nearly covered with a persistent but gently flowing sheet of water. The parallel tracks of the water-filled road were barely visible, but we managed to lurch along them without bogging, and finally arrived at the pub in the late afternoon. It was a rather massive building with a newer, tin-roofed wooden structure tacked onto a larger, older mud-brick original building, both high and dry above the flooded countryside. Simon, the publican had driven out to the west to check on road conditions, and we were greeted by his wife and two small children. We indulged in a few refreshing beers, and were invited to set up camp under the roof of the older mud-brick part of the pub, a welcomed respite from the rain that continued all night. Another customer at the pub was "Dogger George", a true outback character who lived in his ramshackle car and trapped and/or poisoned Dingoes, exchanging their scalps for a Queensland bounty for his lonely living.

At about 10 o'clock that evening two mud-covered men slogged into the pub; one of them we recognized from the Windorah Pub, a traveling accountant who was also on his way to Sandringham, for bookkeeping. The other was an Aboriginal stockman who had hitched a ride. They explained that the accountant's panel van had bogged in the rising floodwater, and they had abandoned it and waded to the pub. With a promise to help them the next morning, we hit the sack. In the morning we accompanied the bookkeeper and the stockman to their vehicle, only to find that it had been left in a deepening creek channel with the windows open, through which part of the creek was flowing, filling it with water, mud and debris. The bookkeeper agreed that it was hopeless, and that we should proceed onward. He was to catch up with us later at Sandringham Station, telling us that the van had taken several days to extricate, and was a complete loss!

Fortunately for us, the morning greeted us with clearing weather, and the storm appeared to have passed. Simon, the publican, helped us refuel from a large barrel, filled with gasoline (petrol) and equipped with a hand pump. We remarked to Simon at the time that he had an unusually long hose that extended from his pump to the nozzle; he merely smiled. We shrugged our shoulders, said our good-byes, and proceeded toward Birdsville. The country all around was greening up fast with new-forming leaves on the perennial plants and trees, and conspicuous germination of annual forbs and grasses. Birdlife was plentiful in the vegetation around the creeks, and only standing pools of water remained where water had been flooding a short time previously. It was apparent to me that, at least in this part of central Australia, nature responds very rapidly to rainfall with renewed growth and vigor. With the weather clearing, lovely new, green vegetation cropping out of the desert, and a couple of cold "tinnies" under our belts, the days ahead looked very bright indeed. Surface water from the previous storms was rapidly evaporating and sinking into the soil, and the dirt track was becoming easily navigable. Creek crossings had also been quite manageable, although we generally walked them before driving. About 10 km west of Betoota we came to a creek with several parallel channels. Although it had been flooded previously, it was no longer carrying water and the bed was damp sand and mud, looking rather sticky. At Betoota, we had been advised to take the longer, wet-weather road to Birdsville, and got to thinking that perhaps we had missed the turnoff, and had wound up on the wrong track. So, we returned to Betoota, where Dogger George agreed to escort us to the right track in his jalopy, only to find that we had been correct all along. With Dogger George's assurances that "she'll be right, mate" as he departed back to Betoota, we reconnoitered the first, sticky channel on foot, finding what appeared to be a firm path.

It was my turn to drive again, so we piled in and proceeded across this channel easily, with the intent of stopping on higher ground to inspect the next channel before crossing. I reasoned, with the first channel being so easy, that surely the second would be no different, and that was my mistake! We started out on nice solid footing, getting about one-third of the way across when suddenly the Landcruiser began to sink vertically, tires spinning, and we churned to a stop. I put it in four-wheel drive, low-range transmission and attempted to back out, only to sink further vertically. I muttered a few rude epithets, and knew we were pretty well mired. We got out of the vehicle, and could see that it was resting in a sea of jelly-like mud, sunk down to the axles with all four wheels. Walking adjacent to the vehicle revealed the creek bed to be reasonably firm until we jumped up and down on it, and then it, too, turned to a jelly-like consistency. Apparently the fine clay-like sediments were completely saturated with water and were thixotrophic, turning readily from a solid to a liquid state with a little agitation.

To lighten the vehicle we unloaded it and then began a trial and error (mostly the latter) process of all the conventional things one does when stuck in mud or sand. We soon learned that the vehicle was sunk too deeply to get a jack under the axle to elevate the tires in hopes of placing something underneath for purchase. Placement of small branches behind the tires merely resulted in the vehicle sinking deeper when put into gear. As it was afternoon and we clearly were making no progress, we decided to set up camp, and wait for inspiration. While Tony, Dione and I viewed our predicament as merely a temporary inconvenience, Mike Waldman became quiet and brooding. That night we had a good dinner, a large campfire, and, stimulated by a liberal amount of Bundaberg rum, had no inspiration and retreated to the sleeping bags. We awoke to a bright clear day and, while the rest of us sought inspiration, Mike became broodier. After breakfast we grabbed the shovel as a substitute for inspiration and began to dig out the tracks behind the vehicle, from the direction I had driven it in. However, as fast as we dug, the clayey ooze flowed in to replace what had been removed. At last, inspiration struck, and we realized that we would have to build a dam around the front and sides of the vehicle to keep the surrounding mud from flowing in, and that then we might have a chance of removing that within which the vehicle rested and in which it was mired.

In the meantime, Mike had gone off to sit under a tree where he was writing a note to his wife. We now understood Mike's sullenness, as it dawned on us that he probably thought we might be stuck in the outback, as had been Burke and Wills. However, we were able to persuade Mike to join us in gathering rocks and pebbles to force into the mud on three sides of the vehicle to form the walls of a dam. Although there were no large rocks around, there were plenty of smaller ones on the sides of the creek. The warm day aided in drying out the mud and we could see that our plan was working. In about four hours a dam was constructed which stopped the flow of mud towards the vehicle; and, with sharing the shovel work under it, we could see that as we removed the muck, the deeper clay was more solid. With the aid of the dam and the drying sun we literally dug down to a firmer, drier layer, sloping it back to the edge of the creek. By mid-afternoon we backed the Landcruiser out of what seemed to be a hopeless situation, and rejoiced. What we really avoided was not a truly desperate situation, but a long, tiring walk back the 10 km to Betoota for help, and the admission there that I had gotten us bogged.

We loaded up the vehicle to make certain that another night was not spent at that creek crossing, and you can be sure that we carefully paced off every inch of the way across, stomping on the creek bed as we went. "No worries, mate!" we made the crossing without a hitch, leaving our stone dam excavation behind as a monument to "Bogger Dick". That night we camped about a half-day's drive east of Birdsville in country similar to where we had been bogged, with alternating creek channels and higher ground. This topography, we were to learn, is typical of the Channel Country between Cooper Creek and the Diamantina River that we were to reach at Birdsville (map 3).

Before arriving at our campsite, we had experienced some difficulty in shifting gears, as if the clutch was not performing properly. Mike Waldman was the only one of us with any mechanical experience and he volunteered to look into the matter in camp. After setting up camp, he started the engine, and reported that the gears were frozen. This sounded serious indeed. Mike crawled under the vehicle with a hammer and began pounding and cursing, and cakes of dry mud began to fly. He crawled out from under, started the engine, shifted into reverse, then low, and took off down the track. It worked, Mike had repaired the vehicle, earning our respect and admiration. Apparently, the clay from the bog had caked around the clutch pedal and cable and had hardened, impeding its action.

The weather remained clear with warm, sunny days and cold, clear nights. An inspection of the terrain between creek channels near the camp revealed small mouse-sized mammal tracks. This observation helped to divert our obsessive attention on four-wheeling along rain-soddened tracks to the much more stimulating adventure of scientific inquiry; but I was reminded, that as so often happens in field work, the fun of discovery is frequently preceded by the drudgery of its pursuit. Discovery of these tracks prompted us to set out 21 live-capture, small mammal traps, baited with rolled oats. Much to our surprise and real joy, one of the traps the next morning held a Fawn Hopping-mouse, *Notomys cervinus*. It was a pregnant female, a lovely blonde-colored, and delicate little rodent with very long, pink ears, large dark eyes, elongate hind limbs and a very long tail for balance while hopping. Except for the longer ears, it resembled superficially a North American Kangaroo Rat, *Dipodomys* spp. I was truly elated, and found myself filled with optimism for collecting additional specimens of this handsome rodent, and for extracting from them some of their secrets of existence. We carefully tucked it into a holding jar with dry sand, bedding, seed and apple for safekeeping.

We then proceeded to Birdsville, nearly without incident. Again, I happened to be driving and it was very thirsty weather. As we drove into sight of the Birdsville Pub I could not take my eyes off it, so intent was I to get us there. As it was Sunday with noon rapidly approaching, we were afraid the pub might close soon, and therefore we could not afford delay. I managed to stay on the dirt track easily, but I failed to note a large pool of water that extended across it, and drove right into it, becoming bogged once again! Now it was my turn for scorn and derision, and it was heaped upon me by all. I was forced to admit to the appropriateness of the name "Bogger Dick", which, in some quarters, I bear to this day. Fortunately, this was only a minor bog, and a little pushing and backing got us out handily. In addition to embarrassment, this latest bit of skilled driving cost me more than one round of beers at the Birdsville Pub, which, as it turned out, wasn't even aware it was Sunday and was eager for customers. Birdsville is in the extreme southwest corner of Queensland, located about 150 km east of the Northern Territory border and 50 km north of the South Australia border.

After refueling at Birdsville, we headed north, along the eastern edge of the Simpson Desert, toward Bedourie just beyond which is the track leading the 56 km northeast to Sandringham Station. As we drove north the country began to look more familiar, reminiscent of parts of the Sonoran and Mojave Deserts with which I grew up in the southwestern United States. Vast stretches of sandy, open country supported sparsely distributed *Acacia* shrubs that resembled very closely in life form the creosote bushes of the North American deserts. Carpets of desert annual flowers had germinated in response to the recent rains, and were beginning to bloom, forming white and yellow fields as they do in the California deserts. However, two other uniquely Australian features reminded me quickly that, while each desert shares similarities with other deserts, each also has its own characteristic features. In Australian deserts, the sandy soil is a deep orange-red, and the Simpson Desert in particular is composed of vast stretches of parallel red sand hills with rust-colored clay pans between. The sand hills are made up of fine, loose wind-blown sand, supporting some perennial grasses and shrubs, but more bare than vegetated. These sand hills course generally from northwest to southeast, each may be several to many kilometers in length, generally less than 200 meters in width, and up to 50 meters in height. Near Birdsville, Bedourie and Sandringham Station the dune density is classified as low, about three parallel dunes per kilometer (Twidale and Wopfner 1990). The clay pans between the sand hills are flat, bare depressions in which runoff water accumulates after rains. Since they are the low points of the landscape and the runoff water has frequently flowed across a fair bit of desert terrain, both clay and dissolved salts are deposited on them, leaving them, when dry, as barren, salty aprons, up to 50-60 meters across, with frequently a sparse growth of halophytic (salt-loving) perennial plants around their margins. Especially common are species of the cosmopolitan Saltbush genus *Atriplex*. Thus, the areas between the sand hills are mosaics of clay pans interspersed with low stands of halophytic vegetation. We were to learn also, as we approached Bedourie and Sandringham that to the east the Simpson Desert is bounded by stony desert or Gibber plain, which interdigitates with the sand hill-clay pan country. These Gibber plains are constituted of open, barren stretches of wind and sand polished rocks and stones which concentrate on the surface as the sandy soils in which they were originally embedded are blown away, to be deposited on the sand hills. Thus, the area consisted, as we proceeded, of three rather different, juxtaposed habitats: sand hills, clay pan, and Gibber plain.

For the last day or two Tony, Dione and I had noticed a rather pungent odor emanating from the interior of the vehicle, and attributed it to a bit of rotting food that had lodged in some hidden cranny. On this particular occasion Dione and Mike were riding in the back and Tony and I were up front when all of a sudden we heard Dione exclaim, "Mike, it's you. You're what bloody smells!" The three of us had bathed periodically in various earthen dams and tanks, filled with runoff rainwater that, even though reddish-brown in color like the surrounding country, was clean and refreshing. Mike had declined to bathe, cleansed in his British dignity, up to that point. With this revelation, and since it was approaching lunchtime, we held a hasty conference. In a generously democratic manner we agreed to stop at the next dam for lunch and a bath, and Mike was advised to bathe or walk. The world was a better place after Mike elected not to walk!

That evening we camped between Birdsville and Bedourie, with a red sand hill on one side and Gibber plain on the other. Again, the 21 small mammal traps were set, baited with rolled oats, and we were delighted to catch two more rodents near the base of the sand hill. One was another species of hopping-mouse, the Spinifex Hopping-mouse (*Notomys alexis*) resembling in my mind North American pocket mice (*Perognathus* spp.). The second was a smaller, more conventionally proportioned mouse, the Sandy Inland Mouse, *Pseudomys (Leggadina) hermannsburgensis,* that resembled the North American Harvest Mouse (*Reithrodontomys*). In two consecutive nights, we had captured one each of three experimentally unknown desert rodent species, and we hadn't yet arrived at Sandringham where they were supposed to abound. Tony's and my excitement levels mounted as we sensed a unique opportunity to study a nearly unknown ecological system.

We then proceeded on to Sandringham Station where we introduced ourselves to the manager, Reg Arthur, and his wife, Marg. Reg eyed us suspiciously, for we were an unlikely lot: an American, a Britisher, and two

Australians, one of whom was a married woman traveling with us, while her husband was back in Melbourne. An additional reason for Reg's suspicions, he told me later, was my connection with Basil Marlow, who had originally told me about Sandringham. Reg and Marg Arthur possess a genuine love of the land and its flora and fauna, and Reg had taken Basil out spotlighting one night to show him some of the native mammals he admired. Among those they saw was a Kultarr (*Antechinomys laniger*), a small dasyurid marsupial that, at the time, was considered rare. While Basil was telling Reg how rare this animal was, he proceeded to dispatch it with a shotgun for the museum collection. Reg was unable to rationalize killing a rare animal, and was understandably upset. His suspicions about us were appeased somewhat when we explained that we hoped to collect live samples of several species of small mammals. These samples would be used back at Monash to study their adaptations to deal with the rigors of a desert existence. We explained that we were more interested in how they made a living than how they looked stuffed or pickled on a museum shelf.

The Arthurs introduced us to Andy Keefe, a young American who coincidentally was staying on the station. It turned out that Andy had been hired by the University of California, Davis, School of Veterinary Medicine to come to Australia to collect live small dasyurid marsupials. These would be sent back to California in hopes of establishing laboratory colonies to be used in cancer research. Andy was a keen and careful observer, was liked and trusted by the Arthurs, and had been very successful in catching the Kultarr and the Kowari (*Dasyuroides byrnei*), another dasyurid. They were caught with traps and with a hoop net at night, thrown from his Minnie Moke, a small, stripped down jeep-like vehicle. Andy also confirmed that the Fawn Hopping-mouse, *Notomys cervinus*, was very abundant, especially on the clay pans. He volunteered to help us catch animals.

Reg reserved his suspicions and showed us a place to camp near the station buildings. He urged us to drink rain water from their tanks filled from a roof-collection system, explaining that their bore (well) water was highly mineralized, and would likely give us intestinal problems for a matter of days until we got used to it. We tried hard not to impose and apparently succeeded, for part way through our stay, the Arthurs presented us with an entire beef tenderloin. This provided lovely grilled filet mignon for two consecutive dinners. One evening, though, Reg came by our camp to see how we were getting along, and asked, "Where's the scruffy one?" Tony was the only one who wasn't there at the time! Another outstanding feature of Sandringham Station is its artesian bore, out behind the homestead, where warm water gushes continually from a pipe at about body temperature. What a place to bathe after a long day; even Mike Waldman liked it!

For three consecutive nights (22-24 August 1966), we set traps and spotlighted mammals at night with Andy Keefe's help. He shared his secret bait recipe with us for dasyurid marsupials (a mixture of peanut butter, bacon bits and bacon fat), and instructed us in throwing a hoop net over an animal held in the beam of a spotlight. We also set traps baited with rolled oats in hopes of catching rodents. These were nearly moonless nights, and to our amazement and delight, we saw many small mammals, both dasyurid marsupials and Fawn Hopping-mice. In three nights of netting and trapping, we caught 18 Kultarr, 6-8 Kowari, 6 Fawn Hopping-mice, and 1 Spinifex Hopping-mouse. Interestingly, nearly all of the dasyurids (Kultarr and Kowari) were caught in traps, one Spinifex Hopping-mouse was trapped, while the Fawn Hopping-mice (see fig. 2) could only be netted. This seemed anomalous to me because the desert rodents I have worked with in North America readily enter traps. With the exception of the one Spinifex Hopping-mouse that was caught in a trap on a red sand hill, all of the Sandringham mammals were trapped or netted on Gibber plain or clay pans (see fig. 3), or the ecotone between them. It is still surprising to me that this seemingly barren gibber supported such a large and varied population of nocturnal mammals. Of the dasyurids we caught, we kept only five each of the Kultarr and the Kowari, turning the others over to Andy Keefe for his colony or releasing them. We kept all of the hopping-mice, and the one Sandy Inland Mouse caught earlier for transport to Melbourne for study.

Figure 2. A Fawn Hopping-mouse from Sandringham Station, 1966.

Figure 3. The clay pan habitat of the Fawn Hopping-mouse on Sandringham Station in 1966. The bare surfaces are the clay pans; a sand hill is in the background.

Because we anticipated that especially the dasyurids might be delicate to transport due to their carnivorous diets, we decided to take the most reliable and fastest route back to Melbourne, and to leave immediately. Following the Arthurs' advice, we elected to skirt the channel country and possible bogs and slow going, by looping around to the northeast on more established stock roads, and then south to Melbourne. Our route was to take us from Sandringham north to Boulia and Mount Isa, then southeast to Winton, Longreach and Barcaldine, then south to Blackall, Charleville, Cunnamulla and Bourke, then retracing our earlier route back to Melbourne (map 2).

Since I had written Ken McGuire, the manager of Marion Downs, a station between Sandringham and Boulia, requesting permission to collect there, we felt obliged to make a courtesy visit. We spent one night at Marion Downs spotlighting, saw one Fawn Hopping-mouse and failed to catch it, but saw nothing else. We explained to Ken the next morning that we hadn't seen much, and that we had to move on to safeguard our captive animals. He was a bit perturbed that we felt Sandringham was a better place, and indicated that our departure was hasty. But depart we did, with the only memorable event on the way back to Melbourne being a breakfast birthday celebration for Dione at a camp amongst the Spinifex clumps east of Mount

Isa. She was greeted with champagne and orange juice purchased on the sly the previous afternoon in Mount Isa. The return to Melbourne was made without further incident or rain in three or four days, losing only one animal (the initially-trapped Fawn Hopping-mouse) in transit. This brought an end to my first marvelous meandering in the bush, whetting my appetite for more.

Upon return to Monash University, our animals were soon under the watchful eye of Lucien Vanderbeldt, the animal caretaker, and I was then put in contact with the late Peter Aitken, at that time Curator of Mammals at the South Australian Museum in Adelaide. Peter had just returned from the vicinity of Yuendumu settlement northwest of Alice Springs and at the southern end of the Tanami Desert in the Northern Territory. There, with Aboriginal help, he had collected living samples of the Spinifex Hopping-mouse (*N. alexis*) and the Sandy Inland Mouse (*Pseudomys hermannsburgensis*) and had returned with these to Adelaide. Since we had managed to bring back only one or two of each of these from Sandringham, insufficient for the study Tony and I had in mind, Peter graciously agreed to loan to us a dozen of each species, and arranged for their air shipment to Melbourne.

Shortly thereafter, I was joined by my first wife Ann, and our children Jennifer and Douglas, then age 10 and 7, who had spent the previous couple of months in Hawaii. We set up housekeeping in an older furnished bungalow in the Melbourne suburb of Canterbury, with the imposing address of 1 Parlington St. I had never lived at a number one before! We settled the children into a nearby elementary school, which was neither demanding nor particularly stimulating. We later enrolled Jennifer in Lauriston, a local private girls school which was much better suited to her abilities; Doug seemed content in his original class. I then turned to the less adventuresome task of experimental science.

I was still hopeful of collecting an experimental sample of Spinifex Pigeons (*Petrophassa plumifera)* and made contact with an amateur ornithologist (a “birdo” in Australian lingo) in Mount Isa, Queensland, about 500 km north of Sandringham. Upon being informed that Spinifex Pigeons were plentiful around The Isa, and since we had a barely adequate sample of Fawn Hopping-mice from Sandringham, Tony and I decided to make a hurried, dual-purpose trip back to Mount Isa and Sandringham.

In mid-October 1966, we flew from Melbourne to Mount Isa via Brisbane, Queensland. At Mount Isa the only rental car available was a conventional Ford sedan rather than the four-wheel drive we had hoped for, promising another meandering adventure. We were rather ill equipped for camping as the Arthurs had invited us to stay in a spare bedroom in their homestead at Sandringham, so we spent the first night at a motel in

Mount Isa. Early the next morning we started south for Sandringham and, after traveling about 100 km, a passing truck pitched a rock from its tyre (tire) at our windscreen (windshield), which fragmented into our laps. We were more startled than annoyed or amused, but returned to Mount Isa and the car rental agency, and were directed to a windscreen repair shop. They assured us it would be repaired by late afternoon, so we arranged for another night at the trusty motel. The windscreen was repaired on schedule, and Tony and I treated ourselves to a seafood dinner at a nearby Greek restaurant, which otherwise was of a rather nondescript nature, or so I thought! We should have known better than to order seafood in the desert, and far from the coast. That night Tony complained of an upset stomach, but mine seemed to be okay.

Early the next morning we started off again for Sandringham with our new windscreen and high hopes. The weather was fine and the road was dry, but dusty, and it was a well-graded and well-traveled track. We should "be right" without four-wheel drive. Tony's upset stomach seemed to have passed in the night along with a lot of gas and several trips to the toilet (loo, or dunny in Australian vernacular). However, shortly after we started south my guts began to churn, and it soon became clear that my lower gut was going to require immediate relief. I felt might crook (sick) and Tony sympathetically advised me that, since he had survived the night, it was unlikely food poisoning, and more likely a temporary affliction like the Barcoo Rot or the Dreaded Lurgie, both of which were common bush maladies. With that he pulled over to the side of the road, handed me the roll of toilet paper we had liberated from the motel, and waited patiently. The trip south to Boulia was a series of starts and stops, and each time I feared that the stop would not be in time. Tony was driving and realized that the deposit on our rental car was at stake, and, reading the level of agony on my face, was able to gauge the distance traveled between stops with great accuracy. We reached Boulia about noon, by which time I was pretty well cleaned out and feeling better, but not well enough to eat anything.

Tony and I had been so impressed with Andy Keefe's hoop-netting technique for capturing hopping-mice that we had brought along netting but had no hoop. We located a welding shop in Boulia and in short order a gentleman welded us a hoop about a meter in diameter from 3/8 inch construction rod (metric and English measures were commonly used in the same sentence in Australia, particularly since the country had only converted to metric that same year, 1966!). Onto the hoop, we could easily sew our netting.

We arrived at Sandringham late the same afternoon, where we were greeted affectionately by Reg and Marg Arthur, whose earlier reservations had apparently been completely dismissed. No sooner had they greeted us when, apologetically, they informed us that they were leaving for Boulia for the next day or two. They explained that Slim Dusty and his traveling country band were performing in Boulia that night and the next, and it was something they couldn't afford to miss. Slim Dusty was Australia's most colorful country singer, especially famous for his song "A Pub with no Beer", and revered by country folks young and old as a legend. As we were so intent upon catching animals, we passed up a rare opportunity of seeing bush people at play, and I still regret it. I made up for the oversight partially some years later when I saw Slim Dusty and his family perform at the Sydney Opera House, but it couldn't have been as good as in the bush at Boulia. The Arthurs showed us to our bedroom, urged us to help ourselves to food in the refrigerator and the meat locker, and to make ourselves at home. They said the station was in our hands, "not to worry", and that they would be back in two or three days. In addition, Andy Keefe was now working for them as a drover (cowpuncher), and volunteered for our use his Minnie Moke, which would make netting animals much easier than in our sedan. Then they were all off for Boulia.

We spent the next three nights netting the Fawn Hopping-mouse, and captured a total of 16 to return to Monash. Most of these animals were collected along the margins of the clay pans, where they seemed to be most numerous. While one of us drove in the open Moke, the other sat on the side of the vehicle, holding the net. When a hopping-mouse was spotted, the driver held it in the headlights and proceeded at a slow speed. The animals appeared to move ahead at whatever speed the vehicle was moving, and they seldom darted off to the sides. This allowed the netter to jump off the Moke, run parallel to it and the rodent, and then fling the net with a circular rotation, much like a Frisbee, with the leading edge up in the air and the trailing edge at ground level. If properly thrown, the animal would be scooped into the net by the trailing edge and then the entire net would fall over it, coming to rest. Although misses were exhausting and numerous, this was really great fun, particularly when we decided to reward ourselves with a cold beer for each successful catch. It was fortunate that we happened to have an esky (ice chest) in the Moke. Within three nights we had collected 16 Fawn Hopping-mice, a full complement we decided, and besides we were about out of beer. Between nights of netting, we spent part of each day exploring in the Moke. On one of these occasions we had taken a track out east of the homestead when, about 3-4 miles out, the Moke came to a sputtering halt. Upon inspection, we discovered that we had run out of fuel. It was a warm spring morning, rapidly becoming hot, and we were without anything to drink. With clouds of bushflies swarming around our sweating faces and torsos, we trudged on foot back to the homestead for much-needed drinks of water from the rainwater tanks. After rehydrating and lunch we drove back to the Moke in our rental car, refueled the Moke and drove both back to the homestead, determined not to make such a silly mistake again.

The Arthurs returned the morning of our last day at Sandringham, in very high spirits after their break in Boulia. Slim Dusty had been wonderful, as had been their time there. They declared that day a holiday, giving us a chance to get to know them a little better. Marg and Reg are genuinely warm, affectionate people who really care about the land and nature. They were proud that they had such a rich small mammal fauna in a region that had been grazed by cattle for many years (at least since 1884).

Marg proudly showed us her new bottled-gas kitchen range and went on to explain that in the early days she had been station cook, using a wood-burning range, and that during the summer the kitchen never got below 40°C (ca 100°F) day or night. With this new range, the kitchen remained much cooler. She also related the tale of how they had celebrated one Christmas, after a long period of drought. Christmas in Australia is during midsummer and it is very hot, especially in central Australia. The drought had been broken after several years by summer monsoonal rains just before Christmas, and their roof water-collection system had filled their storage tanks. They had gone from water rationing where sweet, rainwater was in progressively shorter supply to, suddenly, a water surplus. As Marg explained it, Christmas day was exceptionally hot and their children were home for the holidays from boarding school. Some member of the family accidentally splashed water on another, and the latter retaliated. Before long, the entire family was engaged in a massive water fight, joyfully splashing away a copious commodity that just a few days earlier had been nearly depleted and precious.

This story reflects the extent to which water governs the quality of life in the outback, for both humans and other organisms. To the Arthurs it meant renewed vigor of the land, replenishment of water supplies for their use and that of their livestock, and renewed vegetative growth for livestock grazing and for the entire trophic system of organisms whose continuation and regeneration were ultimately dependent upon periodic rainfall. The Arthurs intuitively understood these complex physical and biological interactions, and they had spontaneously celebrated this replenishment with a water fight.

That last evening Marg prepared a fine dinner of roast beef for all of us, which we ate on a linen tablecloth in their dining room. It was difficult for Tony and me to believe how thoroughly trusting the Arthurs had been in leaving their home and the station at our disposal, and it was difficult to express fully our appreciation and sincere gratitude, which remain to this day. Bright and early the next morning all were off in their various directions: Marg to catch up around the homestead, Reg and Andy to go out mustering (rounding up cattle), and Tony and I on our way back to Mount Isa.

Our trip to the Isa was uneventful, and we stayed at the same motel for the next couple of nights. We avoided the Greek restaurant like the plague, dining at some far less memorable eatery. On the second morning, Tony and our hopping-mice were delivered off to the Mount Isa airstrip, for his flight back to Melbourne. My day was spent meeting with my birdo friend who worked in the mines (copper, silver and lead) at Mount Isa, and finding out where the Spinifex Pigeons might be caught. I was sent south of town to an area of rock outcrops, where I spotted a few individual birds. These are lovely, quail-like birds, superficially similar to the Gambel's Quail of the North American deserts, yet they are columbiform (doves and pigeons) birds. Like quail, they are rather clumsy fliers, preferring to run along the ground, and resorting to flight only as a last resort. The Spinifex Pigeon is distributed broadly throughout north-central Australia, with a disjunct distribution in Western Australia (Frith 1982).

After finding a small waterhole in a rock crevice that was being visited periodically by birds, I reasoned that Spinifex Pigeons would come there to drink, so I enclosed the entire waterhole in aviary mesh with a funnel-shaped entrance, converting the waterhole into a live trap, with water as the bait. I observed numerous Spinifex Pigeons in the vicinity over the next several hours and, even though it was a warm day (in excess of 30°C), they didn't seem interested in my waterhole. A closer inspection revealed that similar waterholes were scattered throughout the habitat, and water was unlikely limiting due to recent rains. Sitting and waiting for birds to come in gave me a chance to think about our progress to date, and where our studies might lead. With complete samples of three species of unstudied Australian desert rodents, and smaller samples of two species of dasyurid marsupials, it was apparent that we had the raw materials for some very interesting and potentially important work. Still, I felt that the initial reason I had come to Australia was to continue my work with desert doves, and that I had to give the Spinifex Pigeons a fair go. By evening, I was discouraged, for none of the many pigeons I saw came near the trap. I decided to come back one more day, and then depart for Melbourne with whatever I caught. That last day was another long, hot day spent crouching in shaded rock shelters used by wallabies. I didn't see them, but their ticks found me, as my ankles will attest. Finally between two and three that afternoon, three

Spinifex Pigeons entered my trap to drink, and were collected. No more came near the trap as darkness approached, and so I returned to Mt. Isa with my three birds.

By the time I had flown back to Melbourne, I had decided that serendipity had led us to the desert rodents, and that we should concentrate our efforts on them. This would represent the first thorough physiological investigation of any native Australian rodent, let alone three species whose ranges overlapped. The end of October 1966 was rapidly approaching and I was already four months into my sabbatical leave, with about six more months to devote to the experimental research before heading back to California. I had full time to devote to our work, and Tony was prepared to make our study his highest research priority. Equipped with our rather meager observations of hopping-mouse habitat, a lot of intuition based upon our previous work, and pretty thorough information on the adaptive physiology of North American desert Kangaroo Rats, the ecological equivalents of our hopping-mice, we embarked upon a rather ambitious research program.

North American Kangaroo Rats employ a combination of behavioral and physiological tricks to thrive under desert conditions, and it was reasonable to assume that hopping-mice would employ some variations of the same tricks. Behaviorally, Kangaroo Rats confine their surface activity to the cooler night-time, and spend the hot desert days underground in cool, humid burrows, conserving water that otherwise might be expended in evaporative cooling and breathing dry air on the surface. Their physiological attributes include the capacity to void extremely concentrated urine with kidneys that extract and put back into their bloodstreams much of the water that otherwise would be lost in excretion; their feces are voided nearly dry, and they have very low rates of evaporative water loss from the lungs and the skin. In addition, Kangaroo Rats eat largely seeds that are rich in carbohydrates with high yields of metabolic water as a byproduct of digestive assimilation, and, when stored in humid burrows, the seeds absorb still more water from the atmosphere and substrate, and then are eaten. The combination of these behavioral and physiological capacities results in freeing Kangaroo Rats from the need to drink water; they are so good at conserving the water that comes from their seed diets (both metabolic and absorbed water) that it generally exceeds the small amount that is lost to the environment, making them independent of other water sources.

We knew that behaviorally hopping-mice were much like Kangaroo Rats, in that they spent the days in underground burrows, were nocturnally active on the surface, and readily ate and thrived on seeds. Completely unknown was their physiology, so we decided to use a “shotgun” approach to explore it, examining key aspects of both their water and energy regulatory systems. Thanks to funding from both the Fulbright and National Science Foundations I was able to transport from California nearly all of my specialized research equipment, to combine with Tony's at Monash, including an oxygen analyzer to measure metabolic rate and an osmometer to measure blood and urine concentrations. Tony had the equipment to measure urea concentrations, the key waste product of mammalian excretion whose concentration in urine is indicative of the amount of water lost in that process. An aquatic biologist at Monash had a nanometer osmometer that would measure fluid concentrations in nearly microscopic samples, which we were welcome to use and which might come in handy (it did!). Lastly, the departmental electrical technician agreed to convert a household refrigerator into a constant temperature chamber, so that we could hold our animals at a predetermined temperature while we made our measurements. So we were all set, and our efforts would be greatly aided by my ability to hire as a research technician Russell Baudinette, who had just received his baccalaureate from the department and agreed to postpone his year of Honours work to join our research team. Subsequently, Russell was to complete his Honours work with Tony and then join me at U.C., Irvine for his Ph.D., before going on to his distinguished career back in Australia, first at the Flinders University of South Australia and then Adelaide University, where he headed his Department and continued his studies of Australian wildlife. Tragically, Russ died of cancer in the spring of 2004. His always cheerful presence is sorely missed by his family, his friends, and the scientific community.

To the conventional measures of body temperature, rate of oxygen consumption (metabolic rate), evaporative water loss (water lost from the lungs and skin, simultaneously), response of body weight and urine concentration to drinking fresh water and while water deprived, we decided to add an additional measure of water regulatory efficiency: the ability to tolerate and use salty water for drinking. This decision was based on our observation that: 1) the Fawn Hopping-mouse occurred on and near clay pans in which saline water accumulated after rains and which they might drink; and 2) that much of the vegetation in their immediate habitat was halophytic and probably with a high salt content, and which might be eaten. So we decide to provide the hopping-mice with various concentrations of sodium chloride (table salt) solutions for drinking, to determine their tolerances and the capacities of their kidneys to eliminate excess salt in the urine, thereby extracting useful fresh water for bodily processes. In addition to applying these protocols to the two species of hopping-mice, we also applied most of them to the Sandy Inland Mouse.

Our studies revealed that, indeed, these Australian desert rodents possessed extraordinary capacities for dealing with their desert existences, and they, like their North American counterparts the Kangaroo Rats, employ combinations of behavioral and physiological mechanisms to ensure their continuing existences in nature. Like Kangaroo Rats, their water needs can be entirely or nearly met by the metabolic water produced during the assimilation of seeds, to the point that they can subsist probably indefinitely without drinking water, provided they can find seeds to ingest. Unlike Kangaroo Rats, however, these Australian desert mice do not have unusually reduced rates of evaporative water loss; instead, for water conservation, they rely more upon extremely efficient kidneys that extract maximal amounts of water as urine is produced, leaving it in an extremely concentrated state while the animals are maintaining good health without drinking water. In fact, the urine osmotic concentrations of water-deprived Spinifex Hopping-mice and Sandy Inland Mice are the highest and second highest, respectively, ever measured in mammals, and that record has stood now for over thirty years. The urine concentrations of water-deprived Fawn Hopping-mice, while less concentrated, are equivalent to those of North American Kangaroo Rats subsisting without drinking water. When the two species of hopping-mice were provided with saline water for drinking, we observed differential responses both in willingness to drink and in resulting urine concentrations. The Fawn Hopping-mouse, inhabitant of the salt-ladened clay pans, readily drank and maintained weight on progressively saltier solutions, up to the equivalent of water twice the saltiness of sea water (1 molar NaCl solution). The Spinifex Hopping-mouse, inhabitant of the nearby sand hills (less salty), voluntarily reduced its drinking of solutions saltier than sea water, relying instead on the metabolic water yielded from its seed diet. Urine concentrations suggested that the kidneys of Fawn Hopping-mice were better at eliminating salts, while those of the Spinifex Hopping-mice were better at eliminating urea, a waste product of seed assimilation. Ecologically these differences make sense, as the Fawn Hopping-mouse is surrounded by a ready supply of salty water from the clay pans after rain and flooding, and from the sap of the surrounding halophytic vegetation the rest of the time. The Spinifex Hopping-mouse, on the other hand, is a sand hill inhabitant, where surface water and halophytic plants are lacking, but where seeds of other plants are likely an abundant source of both energy and metabolic water. Given this scenario, one might expect that the Fawn Hopping-mouse would be less resistant to drinking-water restriction than the Spinifex Hopping-mouse, and that is precisely what we found. The Sandy Inland Mouse, also a sand hill inhabitant, solves its problems in much the same manner as does the Spinifex Hopping-mouse, subsisting on metabolic water from seeds when water in the environment is unavailable. All three species likely eat juicy insects as an additional water source, eliminating excess nitrogen as urea, with their efficient kidneys. Thus these experiments, together with our field observations, allowed us to explain to our satisfactions how these fascinating critters thrive in habitats that seem inhospitable to us.

These initial bush meanderings of "Bogger Dick" provided rich and rewarding experiences in both adventure and science, reminding me that I had learned a little more about life, friends, nature, and the ways of science, but reminding me also that I had really only begun to learn, and that both learning and meandering should be infinite, joyful processes. My appetite for the bush and its secrets had only been whetted; I knew I would be back.

SUGGESTED FURTHER READING:

MacMillen, R.E. and A.K. Lee. 1967. Australian Desert Mice: Independence of Exogenous Water. Science 158: 383-385.

MacMillen, R.E. and A.K. Lee. 1969. Water Metabolism of Australian Hopping Mice. Comparative Biochemistry and Physiology 28: 493-514.

MacMillen, R.E. and A.K. Lee. 1970. Energy Metabolism and Pulmocutaneous Water Loss of Australian Hopping Mice. Comparative Biochemistry and Physiology 35: 355-369.

MacMillen, R.E., R.V. Baudinette and A.K. Lee. 1972. Water Economy and Energy Metabolism of the Sandy Inland Mouse, *Leggadina hermannsburgensis*. Journal of Mammalogy 53: 529-539.

MacMillen, R.E. 1972. Water Economy of Nocturnal Desert Rodents. Symposium of the zoological society of London 31: 147-174.

Chapter 2

RETURN TO SANDRINGHAM, 1974

Sunday, March 31, 1861. - Camp 43, return. Mia Mia Camp. - Plenty of good dry feed, various shrubs, salt bushes, including cotton bushes and some coarse kangaroo grass; water in the hollows on the stony pavement. The neighbouring country chiefly composed of stony rivers and sand-ridges.

William John Wills, near what is now Sandringham Station. From his Diary on the ill-fated Burke and Wills expedition (Anon. 1861)

Since publishing nearly simultaneously our independent work on marsupial standard metabolic rate, Terry Dawson, of the University of New South Wales, and I had kept in close contact. Terry, then Professor and Head of Zoology, had invited me to spend my next sabbatical leave with him for collaborative work. Terry had been working on locomotion energetics of marsupials, measuring rates of oxygen consumption of animals running on a treadmill, and considerable interest was developing on the differences and similarities of locomotion costs between eutherian and marsupial mammals. I had been impressed on my earlier visits to Sandringham Station that two similar sized, morphologically convergent small mammals shared the same habitat at Sandringham, one a eutherian and the other a marsupial. These were the Fawn Hopping-mouse (*Notomys cervinus*, a eutherian, bipedal, seed-eating rodent) and the Kultarr (*Antechinomys laniger*, an insectivorous dasyurid marsupial approaching bipedalism). They would be ideal for such a comparative study of locomotion energetics because the numerous variables that confound interpretations in many studies would be reduced to two for these animals: taxonomic status (eutherian or marsupial) and diet (granivore or insectivore).

I accepted Terry's generous offer to assume a visiting professorship at the University, and departed for Sydney to spend my 1974-75 sabbatical year. Arriving the latter part of August 1974, I spent much of September familiarizing myself with the Sydney area, getting my laboratory equipment functioning, and making contact with Sandringham Station for permission to visit. I learned that the Arthurs had been transferred to Glengyle Station, between Bedourie and Birdsville, and that Norman Clarke was the current manager of Sandringham. Glengyle and Sandringham are both under the ownership of Kidman Holdings of Adelaide, founded by the late Sir Sidney Kidman and former owner of vast tracts throughout the Channel Country. In the manner that I had learned was both polite and customary, I wrote Mr. Clarke that I might expect a response from him by a certain date only if it would be inconvenient for our visit. As he did not respond, we assumed that our visit was acceptable, and began to prepare for another drive to Sandringham Station, this time from Sydney. Terry and I were to be accompanied on the trip by Betty Eidemiller, a recent graduate in Biology from Whitman College, in the state of Washington. Betty had received a special Watson fellowship for independent travel and study for a year in Australia, after which she would return to the U.S. to begin work on a Ph.D. at the University of California, Irvine, under my guidance.

We loaded our supplies into a three-passenger four-wheel drive Toyota Troop-carrier and departed early on 13 October 1974. Terry had a small 12-volt refrigerator/freezer that operated off the vehicle's electrical system, so that we could make ice, keep food frozen and drinks cold, all luxuries to which I soon became accustomed. From Sydney we headed northwest (see map 4) to Nyngan (Terry's home town) in central New South Wales, then on to Bourke, where we stayed the first night. We ate dinner in a Bourke cafe, and then decided to take in the local movies where we saw the American film "Slaughter" and the Australian film "Alvin Purple"; my diary states that this combination was a genuine bit of "Austerica"! The next morning we awoke in our sleeping bags to the raucous cries of myriad Red-tailed Black Cockatoos (*Calyptorhynchus magnificus*), huge, pterodactyl-like parrots. Then, pushing north into Queensland, we did a final shopping for supplies at Charleville, before heading west through the quaint rural towns of Cheepie and Quilpie, to Windorah. All of the roads were sealed now and, as there had been torrential rains throughout central Australia in December 1973 and January 1974, the country was lush by comparison with my previous trip. Wildlife was conspicuous with lizards, snakes, birds and kangaroos all abundant, both alive and as road kills.

The Windorah Pub looked pretty much the same as it did seven years before, and the silent customers might very well have been some of the

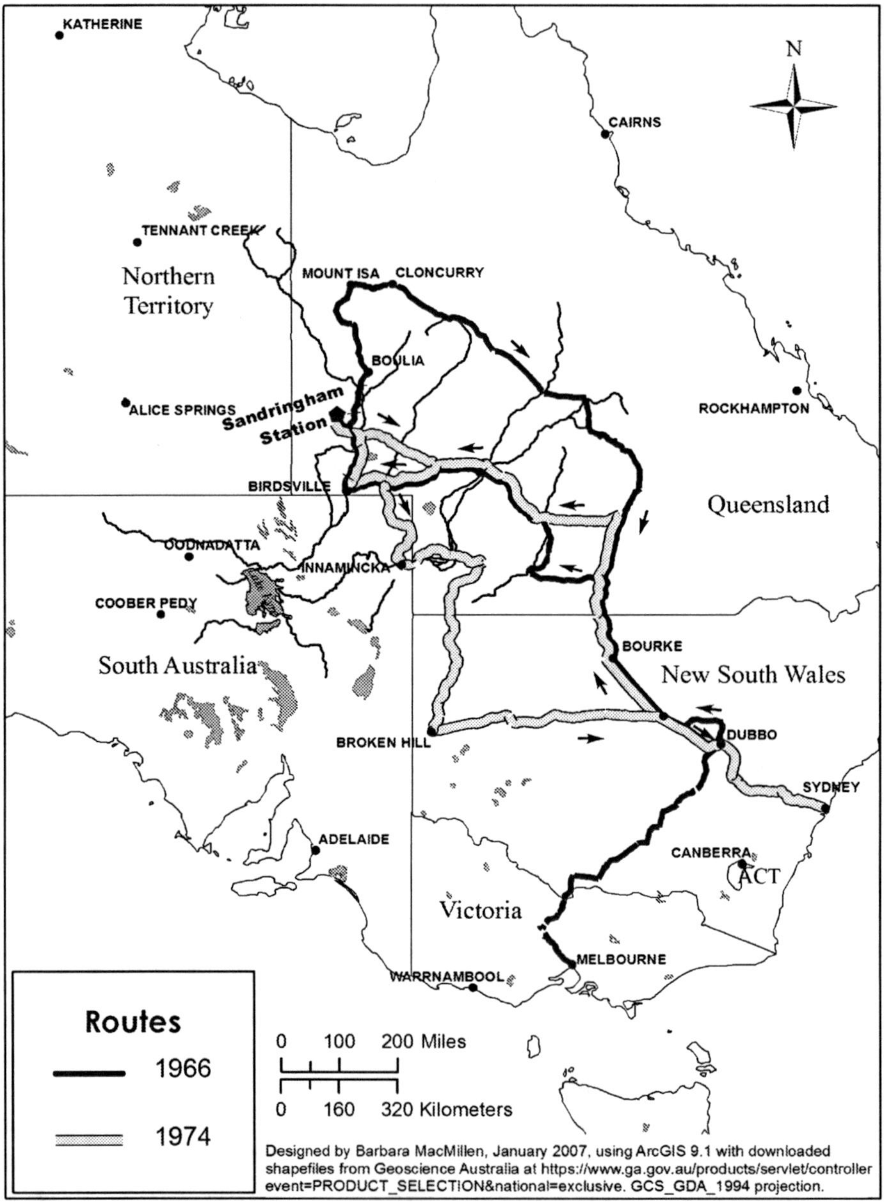

Map 4. Route of our 1974 collecting expedition from Sydney to Sandringham Station and return. Also shown is the earlier 1966 expedition.

same. Their eyebrows lifted somewhat when they saw us, two rather hardened, unshaved characters walking in with a neatly attired young lady. We felt obliged to tell a couple of them our business and, when the publican muttered something about "rat catchers", it was apparent that, once again we had been accepted, but with some curiosity. Windorah is on Cooper Creek which, when it is not flooding, has large pools of permanent water, and several species of fish. Being fishing fanatics, both Terry and I had brought along fishing equipment, hoping to find a place to try our luck. Since fishing is a leisurely sport, and all the people in the pub fit that category pretty well, we decided to ask about fishing in Cooper Creek. When we asked the publican if there was any fishing there, his response was a terse "yup!" So we asked, "What do you use for bait?" He said, "We don't use bait" We asked "How come?" He said, "Too much tucker", which we interpreted to mean that there was so much food in the water that fish couldn't be attracted to bait or lures. Finally, we asked the clincher, "How do you catch them?" He responded, "Net." With a little more prompting the publican went on to explain that he set a gill net in the river, in which the fish became entangled. Terry and I ordered another beer, and the publican seemed to take a liking to us, or at least felt sorry for us. He asked if we wanted to borrow his net and boat, and go fishing. We said, "why not?" Betty was skeptical, but at least game enough to witness this foolishness. The publican left the pub in the hands of one of the customers and directed us to follow him to his house, down the street. There he got out his "boat", which was two truck bonnets (hoods) welded together, oars, and a net, and instructed us to stretch it across the river until we caught fish. He failed to say how long that might take, but he did warn us to set the net upriver, around the bend from the highway bridge, so the fishing inspector wouldn't see us. He didn't mention and we didn't ask anything about legality and fishing licenses.

Off we went upriver about a kilometer to set up camp, set out the net, and then prepare dinner. We set the net across a handsome billabong whose depth we couldn't judge because of the turbid water - nearly all water in central Australia has a milky appearance because of the fine clays in suspension. We managed to pull the net across from one side to the other in the makeshift boat without capsizing, and returned to camp for dinner. We then retreated to the comfort of the pub, fully expecting to pull the net before dark. However, the pub detained us, and it had clearly become too dark to deal with the net until the next morning. After breakfast, Terry stayed on shore holding the near end of the net, and I rowed across to untie the other end. When Terry tried to pull the net toward him it would barely budge. We figured it had snagged on a log. A careful inspection from the boat along the length of the net revealed no log, so I abandoned the boat and swam along the net to the middle. With Terry and Betty pulling from shore and me swimming with the net from the middle, we finally got it moving, but it was apparent that there was something heavy in it. As the net was pulled in it was soon revealed that the heavy something was fish, lots of them! They were Yellowbellies (*Macquaria ambigua*), well over 30 in number, some dead and others alive, and each weighing over a kilogram! It was apparent that we had far more than we could use, so we pulled the net in, releasing the healthy ones, and keeping only those that had been injured in the net. In addition to the Yellowbellies, there were two large Krefft's River Turtles (*Emydura krefftii*) and a blue yabbie (crayfish), which were photographed and released. We cleaned all of the dead fish, filleting and freezing about eight for our use, and hoping to give the remainder to the publican. By about noon we had packed the truck, including boat, oars, net and fish, and returned to the pub. The publican was delighted with the fish, amazed at our luck, and, when we answered his question about how long we had left the net out, his response was "My God, you can catch all you need in an hour!" No wonder we had done so well! We had a beer, thanked him, fueled up and headed toward Betoota. The road remained sealed for another 50 km and then became a well-graded, dry dirt track; nothing like the quagmire Tony Lee and I had negotiated seven years earlier.

As it was thirsty driving, we were happy to arrive at Betoota, where Simon still ran the pub. His wife and children had left him, Dogger George was nowhere in sight, and Simon was glad to have company. He gave us some fresh tomatoes he had grown, and sold us some beers and fuel. My notes say that beer was 50 cents a stubby and fuel was 90 cents a liter, an economy that seemed to favor the right direction. The fuel was delivered from the same barrels, and with the same overly long hose, and with Simon smiling the same smile! It was clouding over and Simon said that the radio reported the likelihood of showers. I was suddenly reminded that it was just west of here, in the direction we were traveling, where we had been mired for over 24 hours on the last trip, so I was determined not to repeat that experience. We drove about seven km west of Betoota and made camp on a creek bank under a large river red gum. That night it rained for about an hour, and we managed to stay dry by crawling into or under the truck, but I was having nightmares about flooding creeks.

By morning, it was clear, warm and windy, the storm having blown over. Standing pools of water were here and there, but there was no evidence of flooding in the creeks and all of the crossings were firm. My fears were unfounded. Between two cattle stations, Durrie and Roseberth, and before reaching Birdsville, we passed through some sand hill country. Water from the last night's rain had collected in a depression at the top of a sand hill, and then had flowed down one side into a large pool alongside the road. The water had receded somewhat, and from the truck we could see tracks in the mud at the pool's edge. We got out for a look and discovered that the pool contained at least a dozen of the same Inland Crabs Tony Lee and I had excavated under similar conditions, about 250 km to the north-east, seven years previously. We collected seven of these crabs, placing them in containers with damp sand into which they quickly buried. It was obvious to me now that these crabs must be widespread throughout the Channel Country, and that they must over-drought underground as adults, and emerge when rainy conditions permit. My curiosity was aroused still further, but after all, I was a vertebrate biologist in pursuit of mammals! We pressed on to Birdsville, arriving at the pub to refuel about 3 PM.

At Birdsville, after the usual libations, we decided it would be appropriate to send postcards to our mentors, so I sent one to George Bartholomew at UCLA, and Terry sent one to Knut Schmidt-Nielsen at Duke University. These gentlemen had been very instrumental in guiding and inspiring our careers in physiological ecology, and the Birdsville Pub seemed like the ideal setting from which to acknowledge this inspiration. After refueling, we headed north toward Sandringham, stopping to camp in sand hill country south of Glengyle Station, now under the management of Reg and Marg Arthur. There we were camped within "cooee" (i.e., shouting distance) of the route used by the Burke and Wills expedition on its way to and back from the Gulf of Carpentaria, about 800 km to the north (see map 1); and they had done this with Camel, horse, and on foot over one hundred years before!

Because of the rain that had fallen earlier in 1974, there was an abundance of forbs and grasses, although peak flowering was long since past. As we traveled northward out of Birdsville, I noticed in areas of densest grass distinctive pathways and occasionally caught a glimpse of yellowish-gray, rat-like animals. Terry and I guessed that they were Long-haired (or Plague) Rats, *Rattus villosissimus*. These native rodents increase rapidly in population, following periods of exceptionally good rains and floods in regions where drought is the normal condition; as more normal conditions return, the populations decline dramatically, becoming restricted to small refuge areas where water and plant food is always available (Strahan 1983). Queensland's interior Channel Country, where we were, is an ideal locality for the Long-haired Rat, with its numerous permanent billabongs and waterholes. That evening at camp, we set 30 live traps, noticing numerous small mammal tracks along the way. Just before crawling into the sack at about 10:00 PM, we checked the traps and released six Long-haired Rats. At 7:00 the next morning, we released 18 more, but caught no other species. The total was 24 animals in 30 traps, an unheard of trapping success for Australia; the Plague Rats were indeed plaguing.

The next morning we dropped by Glengyle Station unannounced to say hello to the Arthurs. Unfortunately, Reg was out mustering (rounding up) cattle and we missed him. Marg was delighted to see us and invited us to come in for tea and cake. We felt rather awkward in our grubby field gear, but she was, as usual, warm and gracious, making us feel completely welcome and at ease. Knowing of our interests in animals, Marg told us about the very large numbers of planigales that retreated to the higher ground around the homestead when it was surrounded by flooding earlier that year. These are tiny, shrew-like, insectivorous dasyurid marsupials and, up until the mid-1970s, were considered quite rare. Marg had captured 28 individuals that squeezed under the door and into the homestead, keeping them in a container for several weeks, after which they all escaped and disappeared. According to her description, and known distributions (Strahan 1983) they were either *Planigale tenuirostris or P. gilesi.* Marg also said that the homestead had been isolated for several weeks during flooding, but that they had anticipated this condition and had left a vehicle on high ground at their airstrip, which could be reached by boat, enabling them to get out if necessary either by air or car. Imagine an inundated country that more typically experiences drought! As we departed for Sandringham, Marg advised us that there had been many improvements there since my last visit.

As we drove north between sand hills toward Bedourie, we noted what had to be high water marks on the sides of the sand hills a good three meters above where we were presently driving. I was beginning to understand how aquatic organisms such as our Inland Crab could have such broad distributions in central Australia: During times of flooding the place was a series of interconnected lakes and drainages, and aquatic animals were free to disperse broadly throughout these waterways. That, however, did not explain how the crabs might persist as adults through long periods of drought.

We fueled up at Bedourie, and while consuming the requisite beers at the pub, I noticed on the bulletin board a posted statement that PeeWee Clarke was the current rum-drinking champion of Bedourie. I wondered if PeeWee Clarke and Norman Clarke, the manager of Sandringham Station, were related. Upon arriving at Sandringham about an hour later we were met by Norman's wife, Diane, who informed us that PeeWee was out mustering; it was then apparent that Norman and PeeWee were related!

Diane had been expecting us and introduced us to their children, Ricky, LeeAnne and Desmond, all of pre-school or elementary school age. We also were introduced to the cook, Clara Reece, an Aboriginal lady who had grown up in nearby Birdsville. We also met Diane's sister Marilyn, who served as governess for the children. It was her duty to help them with their schooling by correspondence and radio instruction through Australia's

"School of the Air". At most remote stations children of elementary school age get their schooling in this manner, but attend boarding schools in larger cities for middle and high school.

The station had improved considerably since my last visit. To the old homestead house and large garage and workshop, had been added a brand-new homestead house where the Clarkes lived. A new bunkhouse with several bedrooms was present. Replacing the old, non-refrigerated meat locker was a large walk-in refrigeration room, with both freezing and refrigeration compartments. And, of all things, the Clarkes had a green lawn! While we had planned to camp and prepare our own meals, we were advised by Diane that we could use the bunkhouse (with hot and cold water, and showers), and would take meals with the family. What a treat! Betty was shown to a bedroom with pink walls, carpets, curtains and bedspread, normally reserved for dignitaries from the ownership, Kidman Holding Company. Terry's and my room was more conventional and quite comfortable, with white walls, linoleum floor, and metal barracks bunks. We presented Diane with most of the fresh fish from Windorah, still frozen, and she was delighted and Clara even more so. We were told we could use the refrigeration room to keep things cold, so our beer and Betty's soft drinks were stashed there. We were also assigned the duty of starting the compressor each morning, and turning it off at night. Sandringham had certainly changed! At least the old homestead house was still used for preparing and eating meals, so that there was something familiar around. Diane informed us also of the daily station meal and refreshment-break schedule. It was Breakfast (8:00 AM), Morning Smoko (10:00 AM), Dinner, what we call lunch (12:30 PM), Afternoon Smoko (3:00 PM), and tea, what we call dinner (6:00 PM). We thanked her, but advised her that we would probably miss most of the smokos and some dinners, as we would be out a lot during the day.

In addition to 95 live traps, we had brought with us a hoop net constructed at the University, and we were anxious to look over the countryside at night, to see what was active. On our previous visits, Tony Lee and I had observed that the Fawn Hopping-mouse (*Notomys cervinus*) was active only during new moon and when the moon was low, on or below the horizon, so we scheduled this trip to coincide with new moon. After tea, we excused ourselves, got in the vehicle and drove out along the station track toward Bedourie for some distance. Nearest the station, we encountered many Long-haired (Plague) Rats (*Rattus villosissimus*) that we didn't attempt to net, but saw nothing else. As we approached the sand hills, however, the rat*s* were left behind, and we began to see Fawn Hopping-mice in increasing numbers, both on gibber and clay pans. With the first sighting, we got out the net, and in one hour we saw 25 or 30 animals and successfully captured 10 of 11 that we tried for, all but one with the net. We were most impressed with Betty's capture of the other with a full-body lunge and bare hand, a technique she had learned with kangaroo rats at Whitman College. In addition, we saw three other small, yellowish critters that quickly disappeared into the nearest vegetation, defying capture. We guessed that they were dunnarts (*Sminthopsis* sp.), small, insectivorous dasyurid marsupials. We saw no Kultarr (*Antechinomys laniger*), the marsupial we were after, but were elated with our success with hopping-mice.

The next five days and nights were busy ones, with setting and checking traps afternoon and mornings, tending captive animals and exploring the remainder of the day, and netting most nights, squeezing in sleep where we could. During that period, a total of 315 traps was set, sampling every habitat type southeast of the station and using a variety of baits. In addition, four nights were spent spotlighting and netting along tracks both southeast and northwest from the homestead. Numerous Fawn Hopping-mice were seen at night along these tracks on both sides of the station and, after capturing 16, we declared that an adequate sample, and stopped trying for more. In the many miles that we spotlighted we saw numerous Long-haired Rats, several Kowari (*Dasyuroides byrnei*), a few dunnarts (*Sminthopsis sp.*), but no Kultarr, the marsupial we were after. This alarmed me because we had seen them commonly while spotlighting in 1966, and Pat Woolley had collected three at Sandringham in 1971 (Woolley 1984).

Over four nights of trapping and 315 trap-sets we caught no Kultarr, nor did we observe any, and I concluded that they were either locally extinct or too rare for practicable capture. We trapped one marsupial species and three rodent species in the abundances as shown in Table 1. Of the 13 Kowari (*D. byrnei*) that were trapped, six were retained for transport back to Sydney, as they might prove interesting substitutes for the Kultarr, even though the Kowari were quadrupedal in locomotion (see fig. 4). I was alarmed to trap the introduced European House Mouse (*Mus musculus*) on both Gibber plain and in sand hills, as we had seen none of these animals in 1966. These animals are known to increase in numbers and geographic range in Australia following periods of rain and flooding (Newsome and Corbett, 1975). It was clear from both trapping results and visual observation that the most abundant rodent at that time was the Long-haired Rat, which was absent in August and October 1966, but which Reg Arthur told us had plagued enormously at Sandringham earlier in 1966 following rains. Their abundant presence in 1974 also followed extensive rainfall and flooding in the Centre, as is typical of the species (ibid.).

Figure 4. A Kowari and the live trap in which it was caught on Sandringham Station, 1974. The substrate is gibber (stony desert).

Table 1. Mammals live trapped on Sandringham Station with 315 traps set over four nights between 19 and 24 October 1974.

Taxon	Common Name (*Species*)	Number Caught
Marsupials	Kowari (*Dasyuroides byrnei*)	13
Rodents	Long-haired Rat (*Rattus villosissimus*)	40
	Sandy Inland Mouse (*Pseudomys hermannsburgensis*)	1
	House Mouse (*Mus musculus*)*	5
Total		59

*Introduced

One night while spotlighting from the truck northwest of the station in sandy, hummocky terrain, Terry spotted a fawn-colored mammal about the size of a jackrabbit, with long floppy ears. It was in the spotlight only momentarily and, while I only got a vague glimpse of it from the corner of my eyes, Terry got a good look at it, and was certain it was a Rabbit-eared Bandicoot or Bilby (*Macrotis lagotis*). These were formerly widely distributed throughout central, south, and western Australia, but now occur in only two or three small disjunct pockets of their former range, one of these in southwestern Queensland (Strahan 1983). They are cited by Aitken (1979) as very rare, but occurring in southwestern Queensland between Betoota and Boulia, rather near to Sandringham. A recent evaluation of the threatened vertebrate species of Australia considers the Bilby to be endangered (Kennedy and Burton 1986), and we believe that there was at least a small population of them in 1974 on Sandringham Station. Terry Dawson is a keen observer, has maintained a small, captive colony of Bilbies at the University of New South Wales, and is quite familiar with the animal. Their continuing presence at Sandringham deserves further investigation.

An evening or two later we returned from setting traps and were pleased to see that PeeWee had returned from mustering, and it was good to meet him at long last. Diane was glad to have him home and anxious for his company. After an excellent dinner of roast beef and conversation, we were contemplating one last spotlighting attempt for Kultarr when PeeWee invited us over to the lounge in the new house for a drink. We didn't think that was very fair to Diane, who hadn't seen her husband for several days, but felt obliged to accept the invitation, just for a short drink. As we sat down PeeWee pulled out an unopened liter of Johnny Walker whisky, pulled out the cork and threw it into the corner of the lounge. At least in Australia this means that you are all supposed to remain until the bottle is empty, so we knew the drink was going to be longer than short. Mustering is very tiring, and we suspected that PeeWee must be buggered (tired). We were more than a little tired ourselves, but then I remembered what I had read about PeeWee's rum drinking prowess at the Bedourie Pub. Betty declined the Johnny Walker, Diane took a small dollop, while PeeWee poured out three water glasses nearly full for the three of us. Terry and I nursed ours along as best we could, but PeeWee was always there to top it up. He proved to be a very interesting, obviously a tough cattleman who knew his business and did it well, as expected of Kidman managers. He wasn't particularly small so we asked about his nickname. He had been working around cattle stations since his early teens when he was smaller than everyone else, so he was called PeeWee, and the name stuck. He worked himself up to a position of respect and responsibility, and was now manager of one of the Kidman cattle stations. We watched the bottle as it emptied, drinking as slowly as we were allowed, and were relieved to see PeeWee pour the last drop. At about the same time he began to slump down in his chair and, putting his glass down, fell asleep. We sat there for a moment, but decided it was terminal, and politely excused ourselves, thanking Diane for the fine dinner and evening. We were informed by Diane the next morning that she left him in the chair, where he had spent the night, but had greeted him the next morning with a cold beer! Needless to say, we didn't go out spotlighting that night, declaring Johnny Walker the victor!

Our daytime activities were spent variously but, after we began accumulating Kowari, at least some time had to be spent each day providing them with fresh meat. We soon learned that they ate Zebra Finches (*Poephila guttata*) voraciously, so we spent a little time each day hunting them with a .22 rifle and shot shells. They were extremely abundant around the station wherever there were waterholes, and one per animal per day provided ample nourishment for weight maintenance if supplemented with drinking water. Diane had told us about permanent waterholes (the Seven-mile, Twelve-mile and Seventeen-mile waterholes) along the track that led to the northwest from the station, so one day we drove out there to explore and get some Zebra Finches. Terry brought along a twelve-gauge shotgun, which I viewed as overkill for the tiny Zebra Finches, but he was thinking of human tucker. In addition to Zebra Finches, we bagged seven Crested Pigeons (*Ocyphaps lophotes*) and three ducks. The evening after PeeWee's near demise, Betty and I volunteered to prepare dinner for everyone, and, with two chooks (chickens) thrown in by Diane, I roasted the fowl stuffed with dressing and basted with white wine and orange juice, while Betty baked bread. It turned out excellently, and Clara beamed, saying it was the best bush tucker she had eaten in ages.

One day while shooting Zebra Finches near the station for Kowari food, Clara came out and asked if I would shoot a crow or two so that she could hang them near the garden to frighten away the other crows that were decimating her vegetables and lettuce. I managed to dispatch two, to her delight. This gave me a chance to ask her a few questions. She told me a little about growing up near Alice Springs, and mentioned enthusiastically that her grandfather had lived the "old way", in the bush. I asked if she had any of his items, and she went into her room and brought out his black basalt ax head, carefully formed and sharpened - a truly magnificent piece of work, and highly prized by Clara. I also asked if she knew the small animals around here, and she nodded. Upon showing her a Fawn Hopping-mouse, I was told that it was a Bushy-tail, for which the Aboriginal name was Wilkinti; a dunnart I showed her was called Lurrarree (with the letter r rolled). I also described the large (ca. 3 ft long) goanna lizard (*Varanus spenceri*) we had caught and then released one cold morning (see fig 5). She said I should have brought it to her, that it was "good tucker!" Clara was also in charge of the station's herd of eight goats, all female, and the station's supply of fresh milk. She informed me one morning that the goats had escaped from their yard and, knowing that we would be out checking traps, asked if we would keep our eyes out for them. Luckily, we spied them at the far end of the horse paddock, about eight miles from the station. After some rather fancy cross-country driving in the Toyota, and one hour later, we were able to herd them back to the station, where Clara drove them into their yard with great relief.

Figure 5. A large goanna caught on Sandringham Station; good "bush tucker", according to Clara Reece.

One last episode at Sandringham Station worth mentioning was the first ever (and probably last!) Sandringham fly-fishing contest. Terry and I had brought along fishing equipment which had gone unused thanks to the special bush techniques taught us by the Windorah Pub keeper. We noted that the pools and small stream that flowed out of the warm-water bore at the station had a large population of small fish, called Spangled Perch (*Leipotherapon unicolor*). These are widespread throughout the Lake Eyre drainage system, commonly occurring in natural bodies of water, cattle tanks and bore outflows, apparently dispersing during periods of high water and flooding (Glover 1990). They resembled small trout and appeared to be feeding in the manner of trout so I proposed that we fish for them. I had brought with me a combination spin-fly rod and a fly reel and rigged it with a small dry fly, and sure enough, they went for it. As Terry and I can be rather competitive, the fishing soon became a contest to see who could catch the largest Spangled Perch, each of us using the outfit alternately until a fish was caught. We each caught four or five, and Terry was ahead with the largest. It was my turn, and I eyed a monster finning near a bank of moss. I carefully wafted the fly to it, dropping it into the water just upstream. I held my breath as the fly floated down toward it, looking like an alien bit of feathers floating down a warm bore outflow in central Australia. However, the fish didn't care and, with mouth agape and gills flaring red, it took the fly. I set the hook, pulling the fish out of the water and through the air with one motion, rather than risk the chance of losing it during a long sporting fight. I held it up to show Terry, a truly magnificent seven-inch Spangled Perch, and then released it. I had won the contest. Such is the life of a rat-catcher in the arid deserts of Oz.

On 25 October 1974, we loaded all of our gear and animals, extended our gratitude to the Clarke's and Clara for the warm hospitality with which we were received, and departed for Sydney. We were quite satisfied with our collecting, in spite of no Kultarr, which had seemingly disappeared locally. We elected to take a rather different, more direct route back (see map 4), which would take us from Bedourie, angling southeast and then south to Betoota, and then south, just inside the South Australian border to Innamincka on Cooper Creek. From Innamincka we would head east back into Queensland, then south, entering New South Wales at the WarriWarri Gate, and onto the Silver City Highway. The WarriWarri Gate closes or opens the Dog Fence, an enormously long, vermin-proof fence extending along much of the northern and western borders of New South Wales, and intended initially to prevent Dingoes (the wild dogs, *Canis familiaris*) from crossing from central Australia into southeastern Australia, and European rabbits (*Oryctolagus cuniculus*) in the opposite direction. Its utility is questionable, as there are Dingoes, European foxes (*Vulpes vulpes*), and rabbits on either side of the fence, although the Dingo is more common north of the fence and the fox more common to the south (Kemper 1990). The Silver City Highway is far less elegant than its name suggests, and is a dirt track that extends from the WarriWarri Gate south through Tibooburra and then to Broken Hill. In 1974, it was well graded, but none of it sealed, south of Tibooburra. From Broken Hill, we would travel east to Sydney, back on sealed roads. Much of this route, between Bedourie and Tibooburra, closely paralleled that of the Burke and Wills expedition of 1860-61.

Not too far out of Bedourie, between Cluny Station and Monkira Station on the Diamantina Developmental Road, Terry saw a very large King Brown Snake (*Pseudechis australis*) cross the road, so we stopped the truck to have a look. In fact, there were two, each about six or seven feet long, slithering along in a very unconcerned manner. As these are very venomous members of the cobra family (Elapidae), we kept our distance. They were not aggressive, and continued moving across the gibber toward some destination. They were the only living snakes we had seen on the trip, which is just as well as most of Australia's snakes are venomous. These snakes had likely benefited from the plague of Long-haired Rats, which they could easily swallow.

A little further along, on the track between Mooraberree Station and Betoota, we came upon two large, red Dingoes that were attacking a downed calf. When the dogs saw us, they ran off. The calf was still alive but weakened from loss of blood and unable to stand. Its face and neck were covered with bites and tears, and both eyes were blinded with punctures. The animal clearly would not survive and was suffering, so we dispatched it with a .22 shot to the head. Betty and I saw another calf on the opposite side of the track, and, on walking toward it, we noticed that it also had bloodstains on one side of its face, with that eye torn. It ran off though, seemingly otherwise healthy. These were calves of at least 100 lb. in weight, and we certainly became believers that Dingoes can kill cattle.

The remainder of the trip was uneventful, although the track between Innamincka and WarriWarri Gate was very rough, the worst we encountered on the trip; luckily it was dry, but twisty, dusty, and full of holes with "bulldust". As we crossed through the WarriWarri Gate and the dog fence we noted that the Queensland side was densely vegetated with excellent regeneration of shrubs and trees, while the New South Wales side was sparsely vegetated with little regeneration. This is correlated with cattle grazing and few rabbits on the Queensland side, and sheep grazing and many rabbits on the New South Wales side.

In Tibooburra, we visited the Family Hotel, a quaint pub whose inside walls are decorated by paintings and drawings of Clifton Pugh and Russell Drysdale, two of Australia's more colorful artists who specialize in outback scenes and people. Barney, the publican, informed us that several years earlier a gathering of artists was held at the pub, and that these gentlemen took it upon themselves to adorn the walls with original murals. No doubt, the beer flowed copiously during that particular symposium of fine art!

That night we stayed at Fowlers Gap Station, between Tibooburra and Broken Hill. The Gap is a sheep station, and both an agricultural and biological research station, owned by the University of New South Wales. Having access to accommodation gave us a chance to tend to our animals, shower, have an indoor meal, and a good night's sleep before the two-day drive to Sydney. We arrived back in Sydney on 29 October, 16 days after our departure from there and 6,136 km (3,600 mi) later. We declared the trip a great success even though we were unable to find the Kultarr (*Antechinomys laniger*) and therefore we would be unable to carry out our proposed comparative cost-of-locomotion study. We arrived back with 16 Fawn Hopping-mice, 6 Kowari, and 6 Inland Crabs (one died en route). Terry and I agreed that we could conduct a very interesting study of the Kowari, and that we would concentrate on locomotion studies of that species (see Chapter 7 for summary). Terry was also interested in doing some work later with the hopping-mice, and I was particularly intrigued with the crabs. So, again, serendipity led the way, to other unanticipated studies, including my pursuit of the remarkable desert crab.

SUGGESTED FURTHER READING:

MacMillen, R.E. and T.J. Dawson. 1986. Energy and Water Metabolism of the Kowari, *Dasyuroides byrnei* (Marsupialia: Dasyuridae), While Resting and Running. Australian Mammalogy 9: 87-95.

Chapter 3

THE REMARKABLE DESERT CRAB, 1975

"They come out of the black holes when it rains."
Customer at the Bourke Pub,
New South Wales, when asked if he knew
the whereabouts of desert crabs

It was December 1974, my sabbatical leave was well underway, and the treadmill that was being constructed for the mammalian locomotion work at the University of New South Wales still was not functional. It was becoming increasingly apparent that I needed to focus my major attention on an alternative project. I had been intrigued with the peculiar Inland Crabs that I had seen on two occasions (in 1966 and 1974) during wet periods in southwestern Queensland, yet had been reluctant to think of them seriously as study subjects since I had never worked with invertebrates before; my thinking was pretty much attuned to birds and mammals. It was fortuitous that Terry and I had brought some crabs back to the University shortly after the arrival there of Peter Greenaway, who had recently completed his Ph.D. in Great Britain at the University of New Castle-On-Tyne, and specialized in studies of crustacean comparative physiology. Peter was very excited about the crabs and about the little I could tell him concerning their natural history. We decided then and there to undertake a collaborative study of these animals [Decapoda: *Austrothelphusa (Holthuisiana) transversa*] if we could find a more accessible collecting locality than remote southwestern Queensland.

Peter and I, and our colleague Mike Augee, had agreed to accompany Terry Dawson and a group of undergraduate zoology students to the annual Christmas Camp at Fowlers Gap Arid-zone Research Station, north of Broken Hill in western New South Wales, where Terry, Betty Eidemiller and I had spent a night in October on our return from Sandringham. At the Christmas Camp, we would spend a week guiding groups of students in a series of desert-related research projects. Unfortunately, the crabs did not occur in the vicinity of Fowlers Gap, so Peter and I each chose separate projects on which to work with the students, Peter's with other aquatic invertebrates and mine with Mike Augee on lizards. However, Fowlers Gap is only a two-day drive to Innamincka, on Cooper Creek in South Australia, and in the known range of the Inland Crab. As soon as the camp was over, during the second week in December 1974, Peter and I elected to head north with the trusty Toyota Troop-carrier and camping gear to Innamincka and Cooper Creek, in search of the Inland Crab. We reasoned that it would be an easy trip since we merely would be retracing the middle part of my earlier return from Sandringham Station, and I knew the road. We also could explore creek channels and waterholes along the way for crabs.

Our initial adventure was early the first day when we spied ahead of us on the road a live Western Brown Snake (*Pseudonaja nuchalis*), another venomous elapid. As I hoped to photograph it, we stopped the vehicle in the middle of the road before reaching it and got out. As the day was rather warm and the vehicle provided the only shade in sight for some distance, the snake apparently took a liking to it and crawled quickly under. I got as close as was safe to peer under, only to see it climb up the inner wall of one of the tires onto the rear axle, and then disappear into the underpinnings. We took dead branches and poked here and there, hoping to dislodge it, but not really knowing what to do if we succeeded. They are very dangerous snakes, and are known to become aggressive when agitated. This snake had every right to be agitated! The snake had simply vanished up under the vehicle. After waiting around for a while for nothing to happen, our only alternative was to move on. Hot as it was, we closed all the vents and windows and proceeded, hoping that the snake would drop off the vehicle onto the road, rather than suddenly emerging in the cab from some unknown hole. We proceeded northward about an hour and the snake did not appear. Our uneasiness was beginning to wane so we decided to stop off at the Malparinka Pub for a beer, to give us the courage to travel on to Tibooburra. The publican at Malparinka was the typical sparse-worded type who uttered "G'day" as we entered and served us each a cold Southwark (a bitter South Australian beer) without uttering another word. I hoped that he might know something about the behavior of brown snakes, and offer some comforting insight into our predicament. I explained to him what had happened and asked if he had ever heard of anything like that. He replied "yup!" and said no more. I asked, "Well, who did it happen to?" He replied, "The postman." "Ah", thought I, "Two words, the conversation is improving!" I asked, "What happened then?" The publican increased his response to a staccato burst of words, "crawled into the cab through the floor boards."

"Well, then what happened?", I asked. "Bit the postman." "My God, this is serious", I thought, and then asked "What happened to the postman then?" The publican looked at me and said matter-of-factly, "He died", and then he turned to wash some glasses. That was the end of the conversation, and even though we thought we detected a sparkle in the publican's eye, it certainly did not relieve our anxieties.

We gingerly got back into the truck and with the windows and vents still closed, drove on up to Tibooburra. Luckily, there were no floorboards, but instead a quite solid floor. As there is some fairly decent creek country around Tibooburra, we decided to stop there and inquire about crabs. We elected to go to the Tibooburra Hotel, across the street from the Family Hotel that Terry and I had visited before, Tibooburra being a two-pub town. We found someone there who claimed he had seen our crabs when he was growing up in central Queensland. He informed us that he used to find "side-walkers" in the creeks after rain, which probably was accurate. However, we could find no one else in Tibooburra who knew, or even cared, about Inland Crabs, and we were viewed with considerable suspicion.

By then, we were beginning to have hope that the snake had actually left the vehicle and, after refueling, headed north out of Tibooburra toward the WarriWarri Gate and Queensland, with windows and vents open. At every waterhole and damp creek channel, we stopped to poke around. We startled a few yabbies and small fish, but saw no evidence of crabs. That night we camped quite a few paces from the truck in case the snake was still within it and decided to depart. The next day brought us back to the rough tracks between WarriWarri Gate and Innamincka, where we passed lots of drying pools of water with yabbies, dead and dying fish, but no crabs. No one knew anything about crabs at Innamincka, where we camped for two nights at a lovely site north of town on the bank of Cooper Creek. Cooper Creek had lots of yabbies and some interesting freshwater shrimp that we caught in traps baited with lamb chops, but a very thorough search revealed no indication of the Inland Crab.

It was obviously time to depart. As we retraced our track from Innamincka, after crossing back to New South Wales north of Tibooburra, I saw a nice, large Bearded Dragon lizard (*Pogona vitticeps*), about a foot long, and stopped. Mike Augee and I had led a student project at Fowlers Gap Station on thermoregulation in lizards just before our departure (see MacMillen, Augee, and Ellis 1989). As I had handled many of these dragons and knew them to be docile, easy to manage creatures, I grabbed this one so that Peter could have a close look. I must have poked my forefinger at it to demonstrate its mouth gape, for, before I knew what was happening, it latched on to the tip of my finger and refused to let go. These lizards have very powerful jaw muscles, and this one was munching down most painfully, right at the base of my fingernail. Peter got out a screwdriver and we managed to pry open the jaws, and extricate a badly bruised finger with little puncture wounds where the small, but sharp teeth pierced the skin and even the nail. My finger throbbed so with pain for about an hour that I had to hold it above my head, while Peter drove. Fortunately, Tibooburra was only a couple of hours down the road where we could apply some cold liquid medication, internally of course! We returned to Fowlers Gap and then to Sydney, no closer to a crab-collecting locality than before, but Peter and I agreed that it had been a great trip, nevertheless.

After our return to Sydney, we were more determined than ever to locate an accessible collecting site for crabs. Peter came across a recent publication (Bishop 1963) that was on the systematics and distribution of Australian freshwater crabs, and that identified all of the known collecting localities of our species, *Austrothelphusa (Holthuisana) transversa*. The most promising of these appeared to be at Rossmore Station, near Bourke, New South Wales, on the Darling River flood plain. Bourke is an easy two-day drive from Sydney, all on sealed roads. A call to the station manager, Bill Stalley, resulted in an invitation to camp and collect on his sheep station. In late January 1975 our expedition to Bourke (or slightly Back O' Bourke) was organized with the two of us, accompanied by Lynn Carpenter, a colleague of mine from U.C., Irvine and also on sabbatical leave, and Harry Taylor, a mate of Peter's from the University of Canterbury, Christchurch, New Zealand. The area around Bourke was in the midst of a mild drought, so that it should be an ideal time to search for crabs during a dry period (it was also midsummer, and quite warm).

While refueling at the Caltex station in Bourke, we inquired of the location of Rossmore Station and Bill Stalley. We were directed about 10 miles east of town, near where the Bogan River enters the Darling, near the foot of a local promontory called Mt. Oxley. At the station homestead we were warmly greeted by Bill Stalley, who said that he had seen "side-walkers" on the property, and directed us to a shearing shed and two dams with water near a Lignum (*Muehlenbeckia cunninghamii*) swamp, now dry. We were welcome to camp there, either outside or in the shed, and spend as much time as we wanted looking for crabs.

This looked to be an ideal site, for it had both permanent water in the dams, and a dry swamp, periodically inundated by flooding of the nearby rivers. The swamp consisted of very fine-grained clays, and supported numerous clumps of Lignum, a polygonaceous shrub with long, sprawling stems and branches, and which flourishes in such places in the dry country (Hall, et al. 1972). Our hopes were high as crabs had been collected here previously by Bishop (1963), and the manager, Bill Stalley, had seen them here. First, we assembled five wire-mesh crab traps baited with chops, which we pitched into the dams in hopes that they held crabs. Then we began exploring around, digging here and there with shovels in a very random manner, not really knowing what to look for. We dug up nothing, and the traps held nothing after an hour or so, but we were not disheartened. Then we decided to explore carefully the dry swamp. It was not long before we found shed carapaces, legs and claws of crabs hung up near the bases of Lignum clumps, likely shed when it was last under water earlier in 1974. We knew that crabs were here, somewhere. As we explored further around the margins of the swamp we began to note vertical holes or burrows, some with dried clay thrown up around the edges, going down into the clay bed of the swamp. These were generally about one to two inches in diameter, about the size of the crabs Terry Dawson and I had collected three months earlier in Queensland. We were getting closer. We decided to take our shovels and excavate some of these burrows, but got nowhere but hot and sweaty. Usually we lost the burrow within the first foot of digging, and careful examination of the removed clay revealed nothing.

As it was past lunch time and very hot, we decided to retreat to the nearest pub in Bourke for lunch, a cold beer or two, a careful think and hope for inspiration. The combination of these actions at the pub led us to the question, "What if we return to the University and report that we knew the crabs were there but we couldn't find them." The answer to that we knew would be ridicule and scorn, particularly from Terry Dawson. This led us to speculate that we would then be asked "Did you try pouring water down those holes to see what would happen?" At about the same time one of us got into a conversation with a local at the pub who, when asked if there were any crabs around here, replied "Bloody mobs of 'em!" When asked where they were, he responded "They come out of the black holes when it rains." We asked, "Where are these black holes?" He responded with a sweeping gesture of his arm, and then got stuck back into his beer. We guessed, by that, he meant all around, and thanked him.

We could tell that it wasn't going to rain any time soon, but maybe we could make it rain by pouring water down the holes we had found, even though they were brown rather than black. We returned to the Lignum swamp and, with the two buckets we had brought, began carrying water from the dams to an area with several holes. We poured water down the holes until they were filled, while Lynn and Harry sat watching them. In a hole that Harry was watching the water began to pulsate, and Harry yelled, "the water's jumping!" As we rushed over, a round shape appeared at the surface. Harry grabbed it, and held in his hand for our gleeful inspection of a genuine "side-walker". It had worked! In short order, nearly each of the holes produced a crab, and we then knew how to catch them. They did indeed come out of the black (or brown) holes when it rained! We systematically marked the holes from which crabs were removed, and formed a bucket brigade to fill more holes. Between 3:00 PM and 7:00 PM that afternoon, we caught a total of 20 crabs, with nearly every hole that was filled with water yielding a crab (see figs. 6, 7. & 8). The crabs emerged at the surface within 30 seconds to 30 minutes after filling, and usually between 10 and 15 minutes. We quickly exhausted the supply of holding containers that we had brought for crabs, and were forced to improvise with empty beer cans that we were first obliged to empty. This introduced the technological breakthrough of "Toohey Tins" for crab storage (Toohey's being a New South Wales beer). With the successful capture of 20 crabs, we decided to turn our attention to setting up camp, preparing dinner, and hitting the sack.

Figure 6. A large female Inland Crab on Rossmore Station; it has just emerged from its "black hole". Photo by Peter Greenaway.

The next morning was spent collecting more crabs with the water-in-the-hole technique, and by noon we had 15 more for a total of 35. We agreed that this was plenty to commence our work. We were surprised though when out of one of the holes emerged a water-holding frog (*Litoria alboguttata*), rather than a crab. These and other closely related frogs are

Figure 7. Habitat of the Inland Crab during the dry season on Rossmore Station. Photo by Peter Greenaway.

Figure 8. Habitat of the Inland Crab during the wet season on Rossmore Station. Photo by Peter Greenaway.

widespread throughout central Australia and, during the dry season, typically retreat underground by burrowing, with a bladderful of dilute urine which can be drawn upon as a water store. In the burrow, they become quiescent and develop an impermeable cocoon of partially shed skin (Lee and Mercer 1967), preventing excessive water loss to the surrounding burrow. This one apparently elected an unoccupied crab burrow for its dry-season retreat.

After lunch, we were determined to excavate some crabs from their burrows, knowing that practically every burrow was occupied. This would enable us to determine something about the depth of the burrows, relative dryness or dampness of the clay, and to extract samples of haemolymph (blood) from crabs during the dry season for assessments of hydration states. By gently working a long Lignum stem into a burrow and carefully digging with a trowel and by hand, following the stem, we found we could excavate all the way down to the crab. Harry, Peter and I each dug out a crab, Peter's being 24 inches from the surface, and mine 20 inches (I didn't record the depth of Harry's). Haemolymph samples were obtained from each by inserting a glass pipette into the base of a walking leg and withdrawing a small amount for later laboratory analysis. As the crabs were excavated, we noted that the top eight inches of clay was dry and hard-packed, but then it became progressively softer and moister with increasing depth down to the crab. We also had noted that virtually all of the burrows we encountered were open to the sky, with clean apertures lacking cobwebs or other obstructions, and we wondered if this indicated temporary departures from the burrow, or at least movement within, even during drought.

Peter's excavated crab was extraordinary as it was a large female with about 100 babies held in a brood pouch between the abdomen and thorax. Most crabs hold their eggs in this pouch until hatching, and then release the hatched, free-swimming larvae into the water before transformation into crabs. Based upon this one observation it seemed likely that a larval stage has been abandoned, or drastically shortened, with crablets likely emerging directly from the eggs. We guessed that the crablets were probably retained in the mother's brood pouch until standing water was available into which they were released. How are the crablets nourished during drought? Another good question to ponder. It was unfortunate that this mother crab was damaged during excavation, and died. The young were retained in water, provided with algae for food and returned to Sydney, where some grew to a respectable size.

Before packing up for our return to Sydney, I conducted a quick survey of crab hole density near our chief collecting site. In an area of about 215 x 12 feet, there were 83 crab holes, equivalent to a density of .03 crabs per square foot. I am uncertain how evenly distributed they were throughout the swamp, but we had chosen one of the densest areas for sampling. Our return to Sydney was without incident and all crabs arrived in good condition in their Toohey Tins. Peter and I were eager to begin our experimental work on these fascinating animals.

By April 1975, Peter and I needed a few more crabs to round out the size-related phase of our study, and we especially needed to make some measurements of burrow microclimates. The two of us left Sydney for Bourke at 6:00 AM on 20 April, and drove pretty much straight through to Rossmore Station, arriving there at about 6:30 PM. It was now early fall and conditions were still dry but much cooler. The next day and night were devoted to measuring the temperatures and humidities of burrow atmospheres at depth intervals of 10 cm from the surface down to the bottom. We focused particularly on five burrows in which temperatures and humidities were measured about every four hours between 11:00 AM one morning and 9:00 AM the next morning. A temperature probe and a humidity probe were each attached to two separate Lignum branches calibrated at 10 cm intervals, and these were inserted sequentially into a burrow for measurement at each depth. The temperature probe was attached to a Bailey portable potentiometer, and the humidity probe to a Yellow Springs Instruments Dew Point Hygrometer, powered with a Honda generator. Although noisy, the system was effective and we got the measurements we were after. The next day, after completing these measurements, we excavated the burrows and confirmed that each was occupied by a crab. In addition, haemolymph samples were obtained from each crab, and soil samples at the bottoms of the burrows were obtained for later analysis.

Between bouts of microclimate measurements, we began filling other crab holes with water, and by evening that first day we had caught 12 crabs. These were mostly very large females, but none with baby crabs. In fact, the one female that Peter excavated on the earlier trip was the only one we found with babies - another mystery. On the second day, we collected additional crabs for a total of 30, which would ensure us the completion of the study without another return to Rossmore. The realities of these kinds of studies is that at some stage the fun-filled field work ends, to be replaced with long hours spent in the laboratory, both day and night. That time had come, and I would be hard-pressed to meet my laboratory commitments prior to my departure back to California. That last evening at Rossmore I had left the four sealed plastic bags of clay from the bottoms of the excavated crab burrows on the ground overnight, where they were subjected to surface temperatures. The next morning each of these sealed bags had copious water droplets inside that had condensed on the plastic surfaces. It was apparent that the clay at the bottoms of these burrows held enough moisture after more than six months of drought to saturate with water vapor the sealed air around it at the prevailing nocturnal surface temperatures, resulting in conspicuous condensation.

Our temperature measurements showed that at night the bottoms of the burrows were warmer than the tops and the surface. We reasoned that if a crab came to the top of its burrow at night and cooled down to surface temperature, it could then retreat to the warmer, very humid burrow bottom, and possibly promote condensation of water vapor as droplets on its colder exterior. We reasoned further that if it could somehow absorb or imbibe this condensed water it could conceivably make up water deficits it had incurred earlier. Thus simply noting condensed water droplets on the inside of a plastic bag and the absence of cobwebs within the burrows provided us with some very interesting speculative thought about a source of water for these crabs during drought. Clearly, too, they are dependent upon burrows in this fine-grained clay, which must have unusual water-holding characteristics.

Therefore, what may have been viewed as a gamble in following a serendipitous series of events, appeared likely to pay off with some very interesting results. The remaining several months of my sabbatical leave were devoted to laboratory work at the School of Zoology, University of New South Wales, with time divided into collaborations with Peter Greenaway on the Inland Crab, and with Terry Dawson on the Kowari. By a series of laboratory manipulations and simulations, Peter and I were able to piece together an internally consistent story that explains the probable secrets to survival of the Inland Crab of Australian deserts.

During periods of flooding of the inland waterways the crabs are rather typical aquatic crustaceans, feeding, mating, and dispersing as gill breathing organisms until flooding ceases and waterways begin to dry up. The crabs then retreat as air breathing organisms to vertical burrows dug into moist but drying clay surfaces to serve as refugia throughout the ensuing dry period. Rather than forming eggs immediately for fertilization, the females apparently store sperm, with egg development and fertilization delayed until another wet season is anticipated. While still in the burrow fertilized eggs are then deposited in the brood pouch where they develop directly into crablets, and where they are retained until the next wet cycle and released into the aquatic environment. During the dry period, the crablets likely subsist on yolk from the eggs. If the drought is prolonged, Peter believes that the mothers eat the crablets, recycling their stored energy and nutrients for another try the next year, fertilizing the next round of eggs with sperm still stored from the last wet cycle.

Our laboratory measurements of the haemolymph (blood) of fully hydrated crabs were identical to those taken from crabs in their burrows, showing that even after several months of drought the burrowed crabs were able to maintain a positive state of hydration. In the laboratory, we were able to show that tiny droplets of water, when sprayed in a fine mist on the crabs, coalesced on their carapaces and ran down grooves into the mouth and gill openings. When this spray was marked with a non-toxic dye, it appeared in their guts, indicating it had been imbibed. Crabs dehydrated in the lab would readily rehydrate in this manner. This convinced us that crabs in nature without access to surface water, if they could promote condensation of water vapor on their carapaces, could readily regain or maintain a positive state of hydration. These lab observations together with hydration states of animals in burrows whose atmospheres were at or near saturation with water vapor during drought, convinced us that vertical movements of crabs in their burrows at appropriate times and temperatures would promote an amount of condensation sufficient to ensure a continuing state of positive hydration. These suspicions were strengthened further by laboratory analyses of clay from the bottoms of burrows, revealing that it had very high water-holding capacities, which could provide moist microenvironments even during prolonged periods of drought; and as confirmed by our field measurements of burrow temperatures and humidities.

Other questions that deserved answering was whether the crabs stored fat during the aquatic phase to fuel their energy needs during drought, and whether they had a special mechanism for conserving this fuel to outlast periods of drought. To answer the second question first, we were able to measure rates of oxygen consumption (an indirect measurement of energy utilization) in fully hydrated, well-fed crabs, and then in the same crabs after several days of starvation and dehydration. In the latter state, with a 16 per cent reduction in body weight, the crabs experienced a 50 per cent reduction in oxygen consumption, thus cutting their energy expenditures in half. This initial experiment did not begin to simulate natural burrow conditions, as the crabs were allowed to dehydrate in ambient laboratory atmospheres. To simulate burrow conditions we constructed artificial burrows by lining the insides of plastic cylinders with damp burrow clay, leaving an air-filled crawl space extending from one end to the other. A crab was placed in each crawl space, and the ends of the cylinders were sealed with plastic mesh to prevent escape. The burrows were then placed in a darkened cabinet with minimal disturbance. We hoped that the crabs would not mind having horizontal burrows in the laboratory, rather than the vertical ones in nature. Periodically over 52 days, each burrow was removed from the cabinet at night, the plastic mesh replaced with rubber stoppers equipped with air ports, and rates of oxygen consumption were measured in an air stream passing through the cylinder and over the crab. Following this, each crab was weighed as well as each cylinder, and both were returned to the cabinet until the next series of measurements. In addition, midway between the periodic measurements of oxygen consumption, we measured temperatures and humidities of the occupied burrows while in the cabinet. This artificial burrow experiment revealed that burrow atmospheres remained saturated with water vapor over the first 30 days of the experiment, simulating measured burrow conditions in nature. During this period, the crabs maintained body weight and presumably hydration state. However, the crabs showed a progressive decline in rates of oxygen consumption during this same period, down to as low as 30 per cent of the starting rate. Thus even before the burrows reached desiccating conditions, the crabs assumed a state of metabolic dormancy, greatly reducing their stored energy utilization. Following 30 days the burrow atmospheres were less than saturated and the crabs lost body weight progressively, but still maintained the lowered rates of oxygen consumption indicative of dormancy through day 52, when the experiment was terminated. This experiment convinced us that the crabs in nature, during drought, undergo physiological dormancy, making their energy stores last until the next wet period. Later investigations by Peter Greenaway answered another question: that crabs, upon entering their burrows at the initiation of drought, have sufficient fat (energy) stores to last until the next wet period, even if that may be more then a year away.

Thus, these truly are remarkable desert crabs that "come out of the black holes when it rains". In parts of their distribution, and in particular in the Channel Country of southwestern Queensland where I first encountered them, these rains and subsequent flooding are very infrequent, often more than a year apart, but they do recur at irregular intervals. There these crabs must stay in their "black holes" more than 95 per cent of their lives, waiting for the next rain.

SUGGESTED FURTHER READING:

Greenaway, P. 1984. Survival Strategies in Desert Crabs. In Arid Australia, edited by H.G. Cogger and E.E. Cameron, 145-152. Sydney: Australian Museum.

Greenaway, P. and R.E. MacMillen. 1978. Salt and Water Balance in the Terrestrial Phase of the Inland Crab *Holthuisiana transversa* Martens (Parathelphusoidea: Sundathelphusidae). Physiological Zoology 51: 217-229.

MacMillen, R.E. and P. Greenaway. 1978. Adjustments of Energy and Water Metabolism to Drought in an Australian Arid-zone Crab. Physiological Zoology 51: 230-240.

Chapter 4

THE HAWKESBURY RIVER AND A BABY IN THE BUSH, 1983

> *The Country is woody low and flat as far inland as we could see and I believe that the soil is generally sandy, in the wood are a variety of very boutifull birds such as Cocatoo's, Lorryquets, Parrots etc. and Crows exactly like those we have in England.*
>
> Lieutenant James Cook, 6 May 1770, from his journal aboard HMS Endeavour near what is now Sydney, N.S.W. (Edwards 1999)

While in the Sydney area during 1974-75 I had made contact with Harry Recher, an expatriate from Stanford University, who is an avian ecologist, then with the Division of Ecology at the Australian Museum. Harry had been conducting field studies of Australian Honeyeater Birds (Family Meliphagidae) in Brisbane Waters National Park, north of Sydney, for many years. These comprise a diverse family of nectar-feeding birds whose ecological attributes converge on those of the Hawaiian Honeycreepers (Family Drepanididae) that Lynn Carpenter and I had been studying for several years, and therefore it made sense for the three of us to compare notes. This led to an invitation for Harry and his family to come to California so that Harry could assume a visiting professorship in our Department at U.C., Irvine, which he accepted. Our work with the Hawaiian Honeycreepers had led me to postulate that arrays of coexisting, nectar-feeding bird species, if aligned in a dominance hierarchy, might have energetic cost outlays that reflected their positions in the hierarchy. That is, the most dominant species would have the greatest access to a continuous energy supply (nectar), and therefore would have energy outlays that could be liberal relative to body mass-related predictions. In contrast, those that were progressively subordinate might be concomitantly more conservative in their outlays due to lesser accessibility of nectar, and their energy requirements might be reduced below conventional predictions. An earlier test of this idea with a three-species array of Hawaiian Honeycreepers proved reasonably robust (MacMillen 1981). While at Irvine, Harry agreed that his Honeyeater community would be ideal for testing this model further; his work demonstrated that commonly five or more species coexist (and Ford and Paton 1977). In addition, the literature suggested that coexisting Honeyeaters may be aligned along a size-related dominance hierarchy, and that a broad size differential exists among Honeyeater species (Ford 1979). This sounded ideal to me, and so we began arranging for my next sabbatical leave (1983) to take place in the Sydney area, affiliated with Harry Recher and the Australian Museum.

My wife Barbara and I were married in 1980, and she came equipped with two sub-teen sons from her former marriage, Chris and Steve, so that the logistics for another year in Australia would have to be carefully worked out. As the research would include both experimental and field work, it would be important to have easy access both to a laboratory and field site, and also an accessible school for the boys. These logistical complexities were simplified when Harry and his wife Judy arranged for us to rent a fully furnished house, three doors removed from theirs, on Dangar Island in the Hawkesbury River, north of Sydney. As my lab equipment was portable, it could readily be shipped and made operable as long as I had a power supply. Harry informed us that our house had a separate boathouse, with power, that could serve as a lab. Further, Dangar Island was just a short boat ride across the water to Harry's major field study site, in Brisbane Waters National Park. As I had visited the Recher house in 1974-75, this all seemed possible, and I assured Barbara that life there would be carefree and easy. Schooling was no problem as the boys could go to a high school in nearby Hornsby, reached by a short ferry ride from Dangar Island to the mainland, followed by a train ride to Hornsby. I was concerned, however, about my Museum commitments in Sydney, as Harry had arranged for a Museum Fellowship and appointment as Visiting Curator of Ecology. My anxieties were relieved though when Harry told me that, since the Museum had no Ecology collections, my commitment was where Ecology occurs, in the field, and that I could come and go as I pleased in conducting my research.

It all sounded too ideal to believe, but nevertheless we packed and shipped our gear, and got on the airplane bound for Sydney, via Tahiti and New Zealand, at the end of December 1982. I was anxious to show Barbara what meandering in the bush was all about. On the way we had a lovely, relaxing sojourn on the Tahitian island of Huahine, followed by a trek around New Zealand before flying on to Oz. Somewhere along the way, Barbara surprised me with the news that she was pregnant with our baby, which promised an interesting new spin on our sojourn in the Bush.

Barbara's Description of Life on Dangar Island

I had never been to Australia, or lived outside the USA, and certainly, I had never lived on a small island that had walking paths instead of roads, but that was exactly where we were headed. In fact, before I started dating Dick I had done very little traveling, camping, or meandering. During the previous five years, Dick had taught Chris, Steve, and me how to explore and really appreciate everything that surrounds us. So, we were off on a new adventure and I'll relate some of the things I found to be different from my life back home.

On a hot January day, we were met at the Sydney airport by Harry Recher and Lynn Day, a former graduate student of Dick's at UCI. All of our bags and baggage was loaded into two vehicles and we were transported to Brooklyn, a small town about 40 miles north of Sydney. In Brooklyn, we boarded Harry's boat, a 12 ft. skiff, and across the Hawkesbury River we went. Harry instructed us to take off our shoes and roll up our pant legs as the tide was out and we would have to walk across the mud flat. Sure enough, we sank about 6 inches with each step. When we reached shore, we were escorted to the hose bib on a large rainwater-catching tank to wash off our feet, and then up to Harry and Judy's house for lunch. After lunch, Judy showed us to the house that we would call home for the next 11 months. When the tide had come in, Harry and Dick went back for all our bags and we began to settle in.

Dangar Island is a small island, approximately the size of eight city blocks, that is only accessible by private boat, or a small passenger ferry (approximately 20 passengers) that makes about eight round trips daily between Brooklyn and the island. There are about 200 houses on the island and one vehicle, the fire truck. More than half of the houses are used as weekend retreats, but the rest are permanent residences. Our house was an older one that had been left to a son and daughter upon their mother's death. Most of the household items were still in the house so it was ready just to move into, or so it seemed. Judy presented us with the basic necessities for breakfast the next morning and invited us to join them for dinner. Chris and Steve spent the rest of the afternoon getting reacquainted with the Recher girls, Jennifer and Myela; Jennifer being Chris' age and Myela a year or two younger than Steve (they had gotten acquainted when the Rechers were visiting UCI). Just as the sun was setting behind the island, Judy fixed up two trays of crumbled bread soaked in sugar-water and handed them to Chris and Steve to take out onto their deck. Before they knew what was happening, there were Rainbow Lorikeets (*Trichoglossus haematodus*) landing all over their arms and heads, coming in to feed on the sugar-soaked bread. What an introduction to the birds on the island! Later on, we were to have many different species of birds around; we had one Kookaburra (*Dacelo gigas*) that adopted us and that would sit on the clothesline every morning, waiting for a handout of bacon rind or some other tasty morsel.

The next day it was off to stock the house with groceries, etc. We needed to leave early as the tide would be high and going out, which meant, hopefully, we would be returning as it was coming in again and low tide and the mud would be avoided. The Rechers took us to Brooklyn by boat and then a half hour's drive to Hornsby to shop. We had to set up a bank account, arrange for telephone service, and buy food. Buying the food meant we went to the butcher, produce stall or stalls, bakery, deli and market. At the butcher, we found that cuts of meat were called by different names than we were used to so we spent much time in discussion with the butcher figuring out what cuts we wanted. Some fruits and vegetables also had different names: cantaloupes were called rockmelons, papayas were pawpaws, and bell peppers were capsicums. The array of goods at the bakery was wonderful. Different products at the market took a matter of time to determine which we liked best, like which tomato sauce didn't have sugar in it. At the deli, we bought bacon with the rind on, not smoked as much as our U.S. bacon, and with the small tenderloin still attached. Also, at the deli we bought lunch meat called silverside, which we know as corned beef, and many different cuts of ham. Since the metric system is used, we had to convert how much we wanted into kilograms. Now, with supplies for at least a few days, we returned to Brooklyn, loaded everything in the Rechers' boat and returned to Dangar Island. We hauled our things across the mud (the tide was still out), up the hill to our house, and put everything away. We discovered how small the refrigerator was, and that the freezing compartment was tiny.

During the next few days, we needed to enroll Chris and Steve in school and purchase a vehicle. By telephone, we found out that it would be necessary to take Chris and Steve into Sydney for testing before they could be enrolled in school; so we made an appointment for the next testing session the following week. Our arrival was during school holidays and they would still be able to start school at the beginning of the next term.

The next priority was to purchase a vehicle, so we rode into town with Harry on Monday morning, as he would be returning the museum vehicle he had used to meet us at the airport. Chris and Steve elected to stay and enjoy the company of the other kids on the island rather than trekking around Sydney looking at cars. We looked and looked and decided on a 1983 Nissan Prairie that we thought would meet our needs. The Prairie was a new model and we would have to wait about a week for delivery. Well, we felt good that we were getting all of our tasks completed and would be quite settled soon. Dick and I found the train station and managed to get the correct train back to Brooklyn.

By the next day, we needed to restock our food supply, and this would have to be done by boat and train. We had arranged for use of a small skiff which we could anchor in front of our house, but in Brooklyn we would have to tie it to a partially submerged public dock, as there was no specified docking facility. The Rechers rented dock space from a local boat repairman, but all the spaces there were taken. We quickly learned about tide fluctuations, and the need to keep your boat in the water by moving it every several hours if you want to use it in the near future. If we wanted to catch a morning train, we had to have the boat in the proper position so there was still water under it. This was our first departure unaided by the Rechers, and our boat was on the mud. There we were, all four of us, clean and dressed for town and sinking up to our ankles trying to push this boat across 50 feet of sticky mud. By the time we finally had the boat in the water and crossed over to Brooklyn, we had plenty of time to clean up while we waited for the next train, having missed the one we set out to catch. I approached the cashier at the train station and asked for round trip tickets for two children and two adults to Hornsby. He said, "You want return tickets?" I explained we wanted tickets to go both directions; he informed me that was a return ticket. We had a pleasant ride into Hornsby and as we departed the train, Dick pitched our tickets for the trip to Hornsby into a rubbish bin (trash can), as they were not collected on the train. When we reached the top of the ramp, there was a gate where we were supposed to surrender our tickets. Dick explained that since they were not collected on the train he had pitched them, but he would go back and search for them if that was necessary. The woman at the gate looked at him rather curiously and let us through.

Well, we had made it to Hornsby and needed all of our hands and arms to carry what we had purchased. We returned to Brooklyn by train, managed to get all of our packages and us into the boat and cross the river. The tide had come in and we were able to motor right up onto the sandy beach in front of our house. We carried all our goods up the hill and then put everything away. This had taken all day and we were all exhausted. Not only was it hot and sticky, but also we had to search for everything. Each trip got a little easier, but we were looking forward to getting our car so we could buy more than we could carry in our arms. At this rate, we were spending a lot of time just keeping food in the house.

The following week we took a train into Sydney (by this time we were well acquainted with the trains), and walked across town to the Education Department. Chris and Steve were given a 15-minute test and placed in the middle level of their respective grades. Each grade had at least seven levels (A, B, C, D, E, F, G), and they were both placed in level D. We were then informed that they should go to Barranjoey Head High School, which we knew was not where the other boys on the island went to school. We were shown on a map that Barranjoey was the closest school to Dangar Island, but we then pointed out that it was the closest if you could fly or take a boat, but by train or bus it would take them half the day to get there. We finally convinced the clerks that Chris and Steve should go to Asquith Boys High School near Hornsby, where all of the other boys on the island went to school, and could be reached easily by train from Brooklyn. We came away glad to have gotten the boys enrolled, but apprehensive of the system.

Now that they were enrolled, they would need school uniforms; so we stopped in Hornsby on the return trip and purchased "summer" uniforms which consisted of gray shorts, gray short-sleeved cotton shirts, gray knee socks, and desert boots (like Hushpuppies). The boys were anxious to start school and did so the following week. They would take the ferry from Dangar Island at 7:30 AM and catch the train in Brooklyn at 8:00 AM. We were able to purchase special school passes for both the ferry and the train. There was a train station at Asquith and it was only a short walk from the station to school. One afternoon each week they would travel by train to another school for sports competition. Chris and Steve were highly recruited for the baseball and basketball teams; it was assumed they would excel at both since these were American sports. They also played on the Brooklyn football (soccer) team, and we were surprised they had more experience playing soccer than the other team members. Many of our Saturdays were spent driving or traveling by train to soccer matches.

The following week we received a phone call from the car salesman informing us that it would take a little longer to get our car, but "No worries, mate!" After waiting several weeks, we were informed that the vehicle was on the dock, but no one knew when it would be released, as the customs officers were out on strike. Dick and I decided that we needed a vehicle, and there was no telling how long the strike might continue, so we took the train to Hornsby to look at car dealers there. We found a new 1982 Subaru station wagon, which they were anxious to sell now that it was 1983. It had 4-wheel drive and, although not as roomy as the Prairie, probably would serve our needs. We agreed on a price, gave the salesman a deposit, walked over to the RAA (Royal Automobile Association) to arranged insurance, purchased a cashier's check at the bank, had lunch, walked back to the dealer, and drove off in our new car. It was a bright yellow Subaru, so it quickly took on the name "Yeller'roo ". We still had time to go shopping, so we filled it with groceries and managed to find our way back to Brooklyn while driving on the opposite side of the road. Back on the island, we had to describe it to Chris and Steve so they could find it parked in Brooklyn the next morning, on their way to school. Later on, we purchased a roof rack that enabled us to haul the net poles for collecting birds as well as our camping gear on various meanderings.

Now that we had wheels we could shop less frequently, but we still had limited space for storing meat, so we arranged for loan of a small freezer from the museum, appropriately named a Tucker Box. Dick and Harry loaded it in the Yeller'roo, and then used Harry's boat (it was larger than ours) to bring it across the river. With this addition, we were able to take a trip to town about every two weeks, and purchase milk and bread from the small shop that was on the island by the ferry landing. This shop was quite small and served as the post office, so we made daily treks across the island to collect our mail and pick up other things we might need. The island was serviced by a passenger ferry that ran between Dangar Island and Brooklyn about eight times a day, between 7:30 AM and 6:30 PM. There were no cars on Dangar Island except for the fire truck, which doubled as the trash (garbage) truck for the inland residents. Our trash was picked up from our beach by barge.

The house we rented was typical of many of the houses on Dangar Island. It was perched on brick pillars at the top of a double block (lot) that had quite a steep grassy slope down to a flat grassy area, next to a narrow sandy beach (see fig. 9). We entered from a slightly raised porch that faced

Inland surrounded by a large yard. The kitchen was very colorful with yellow walls, red countertop, gray floor and cupboard doors, and white ceiling. The lounge room (living room) had several Danish style chairs, a small electric bar heater, and a TV; and generally had a very dreary appearance. The boys shared a nice sized room and we had an equally large bedroom, with dressers and small freestanding wardrobes. Both bedrooms, as well as the lounge room, were off a large enclosed verandah, overlooking the river. There was one additional room with a bed in it, but there was barely walking room to get to the bed. There was also an extra bed on the verandah.

Figure 9. Our home on Dangar Island, overlooking the Hawkesbury River, 1983. The beach is in the foreground, and the boathouse (research lab) is at lower right.

The toilet was a flush toilet, but it was located outside in a small separate shed across the walkway that ran parallel to the porch, and was in direct view from the dining room. There was a bathroom inside the house with a bathtub and sink. Off the kitchen, there was a small room with a laundry tub and a semi-automatic washing machine. I was not sure what a semi-automatic machine was, but soon found out. I turned on the water at the valve until the machine was full, and then turned the water off. I then turned a dial to "wash" and it wriggled and jiggled the entire house. When I thought it had managed to get the clothes clean or I couldn't stand the shaking and noise any longer, I would turn the dial to "spin" and it would spin the drum and drain the water. Then in order to rinse the clothes I would go through the same process. Then I discovered how to act like an automatic machine and stand by the machine while it was on "spin" and turn the water on and off until the water coming out of the machine was not soapy. Needless to say, washday was just that, an entire day spent doing the wash. We had a nice large umbrella-like outdoor clothesline, and the only trick was to get the wash hung out, dried, and brought in before it rained, which on numerous occasions we failed to accomplish. Sydney has weather very similar to Seattle, and there were times when our clothes hung on lines strung up on the porch for days in order to get dry. There were other occasions when we had lines strung up in the lounge room with the heater and fan going in order to try to get clothes dry because we were leaving on a trip, or school uniforms were needed. We purchased two single-size foam mattresses with covers that we stacked on top of each other to sit on in the lounge room, which also served as beds for visitors. We also added a pot bellied stove when it started getting cold.

Weeks later, when we finally received our shipment from the States, we decided it would be fun to have an "Open House" to meet some of our neighbors on the island as well as get together with some of Dick's acquaintances in the Sydney area. The arrival of the shipment was necessary because, in addition to Dick's research equipment, it contained the ingredients for several Mexican dishes which we wanted to make and share.

One of the guests at this Sunday afternoon party was an island resident named Jamie Turner, who had a baby boy who had been born on Dangar Island about a month before our arrival. We had a lengthy conversation about her delivery on the island with help from a midwife who also lived there. Soon after this conversation, it was apparent that I would be having a baby while living on the island, and it was exciting to me to contemplate delivering this child at home with all of our family present. I had had no problems with either of my other deliveries, they were short and sweet, and the thought of having to spend a week in the hospital, which was the normal stay in Australia, was enough to convince me that I did want to have our baby at home. Dick and I had not planned to have any more children, but we were ecstatic at the prospect now, and beamed with joy. I visited an obstetrician in Hornsby who cautioned me on all the problems that might occur during a home birth, and encouraged me to have the baby in the hospital. I saw him regularly through the pregnancy.

Life on Dangar Island became easier as we became more and more adapted to island living; shifting the boat so it was always in the water just became a way of life. We built aviaries on the verandah for housing the birds we would collect, and turned the boathouse into a working laboratory (see fig. 9). The verandah also gave us a place from which to feed the Lorikeets each evening, but then they also started coming each morning just at sunrise and sitting on the edge of the roof screeching for food. When it was warm and we had the verandah windows open, some would come inside and start flying around and we would have to get up and help get them back out the windows.

Once the lab was ready to go and we had birds on which to make measurements, life on the island took on a more usual structure. Chris and Steve were off every morning around 7:15 AM to walk across the island to catch the ferry to Brooklyn, where they would catch the train to Asquith and on to school. Dick and I would decide which birds were to be measured and at what temperatures that day. Dick would usually get the constant temperature box brought to the correct setting while I would clean up the kitchen. I would usually make the morning measurements while he would work up data, or be calculating our strategy for the next collecting adventure, etc. By the time the birds were weighed, put in chambers to get used to the temperature, the measurements were made, and the birds weighed again, it was time for lunch. While we were eating lunch, the constant temperature box would be stabilizing at the next temperature, or a new set of birds would be getting used to the temperature. After lunch, the next set of measurements would be made. Between 3:30 and 4 PM, Chris and Steve would arrive home and dinner preparation would begin, and so life went on.

On the weekends, we would relax but usually enjoy being on the island, as that is when the weekenders would come and Chris and Steve had made some good friends with some of the kids that would come. They would play Cricket or Rugby, or sometimes they would borrow Harry Recher's crab nets and go out in the boat and put the nets out. They would often catch a dinner's-worth of Blue Swimmers, and one time they came in with a large Mud Crab. The first time they caught an octopus, they let it go with much struggling to get it out of the net, and when they told Harry, he said, "Why did you let it go? They are good to eat!" The next time they caught an octopus they hurried in with it still caught up in the net so Harry could untangle it for his dinner.

Chris and Steve were not challenged by textbooks or teachers, but learned a lot about sociology and different ways of life. We tried to get them moved into a higher level, but were told that couldn't be done until they received their first evaluations, which would be months later. After the school holidays in April, it was mandatory that the students change from their "summer" uniform to their "winter" uniform, which consisted of long-sleeved cotton shirt with tie, long pants, and sweater (optional). It didn't matter if it was a hot day or not, the winter uniform must be worn.

One afternoon they arrived home as usual on the 3:30 PM ferry; it was hot so they changed into swimming trunks and were in the water. We heard on the radio that a "change" was on its way; it had hit Wollongong south of Sydney, and could be seen approaching Sydney. A "change" is a change in weather, usually from fine to a storm. (If we were in Hornsby shopping and heard that a "change" had hit Sydney, we knew from experience we had to leave immediately in order to get to Brooklyn and across the river before it hit Dangar Island). In about half an hour after they arrived home, we could see white caps forming on the river toward Brooklyn, and in a few minutes we had large waves rolling into our beach. Chris and Steve got out the borrowed surfboards and spent the rest of the afternoon surfing in their own front yard.

Being dependent on boat travel to and from the island made traveling difficult at times, especially at night, so doing ordinary things like going out for dinner or a movie or visiting friends off the island was just not feasible. One night that we did venture into town, upon returning to Brooklyn, we found our boat trapped under the dock and half filled with water that we had to bail out so we could get home. Toward the end of our stay, a new dock facility was built in Brooklyn and we had our own space, which made it a lot easier to come and go, especially after Ian was born.

The last visit to the obstetrician was on a Friday afternoon in late September, and he said he thought I would deliver within the next 48 hours. He was about 9 hours off, as Ian was born about 12:30 AM on Monday morning, October 2nd. The midwife lived just across the path from our house. We alerted her about 10 PM Sunday night that we would probably need her sometime that night as I had been having some contractions that afternoon and evening. We also alerted our other friends who wanted to be present, and about midnight made the first call. Fortunately, Monday was a holiday (Labor Day of all things!), so we enjoyed Ian's birth, first bath, and opened the champagne. The teenage boys were not very interested, and were kept busy in the kitchen boiling water (for tea to drink). Everyone left by 2 AM and we called family and told them Ian was born tomorrow, as it was still October 1st in the US. To many Dangar Island would not be considered "bush", nonetheless most of our family and the obstetrician thought having a baby born at home on an island in a river was having a baby in the bush.

The island was a great place to walk, as there were paths around the island and across the top. After Ian was born and most of the laboratory work was completed, I would usually put him in a Snuggli in the afternoon and walk to the shop to collect our mail and buy milk or bread.

For many years, a party was held for Dick's department at UCI, called the Desert Extravaganza, at his parents' cottage in Joshua Tree, CA, usually the first weekend in November. We decided it would be appropriate to have an Island Extravaganza, and invite friends to come and camp on our property and spit roast a couple of lambs. We sent out invitations and arranged for a Bush Band. Many of the people attending had also been to at least one Desert Extravaganza, so they knew what they were in for, but it was a new concept to a lot of others. None-the-less, about 50 people came, bringing tents, etc. over on the ferry; the Terry Dawson family arrived by small sail-boat that they had sailed from Brooklyn. We had erected a volleyball court on the flat part of the lawn in front of our house down by the water, and introduced the sport to the island. Some of the guests had played before, and it was a great success. Everyone pitched in with preparing food for dinner and breakfast, as I was also busy with Ian who was only a month old. This was a good way to say good-bye to all our friends new and old, as we would be leaving the end of November.

NOW, DICK'S TURN AGAIN...

The Honeyeater energetic study that Barbara and I would undertake involved both field and laboratory components, which needed to be initiated as early as possible so that the flaws could be identified and corrected. Fortunately, I had set up a similar portable laboratory earlier for the measurement of oxygen consumption of Hawaiian Honeycreepers in a cottage at Volcanoes National Park on the island of Hawaii, so I knew pretty much what to expect. Before our departure from the States, I had arranged for my good friends Tony Hulbert and Hugh Spencer of the University of Wollongong, south of Sydney, to fabricate a temperature-control box from a household refrigerator, and I had shipped all of the remaining necessary supplies and equipment from California. The Museum was willing to post a bond against my equipment, guaranteeing its return to California, so there was no hitch in getting it through customs, including the canned green chiles and Mexican salsas essential for careful experimental work. The only hitch was one I had nearly forgotten from my last trip to Sydney, but I was quick to be reminded. That was the fact that it takes three weeks for a shipment of research equipment and personal effects to be transported by ship from Los Angeles Harbor to Sydney Harbor, but it takes four weeks for the same shipment to go from the docks of Sydney Harbor to the customs release point, only a short distance away. Don't ask me why, but it's a twice-proven fact. At any rate, it was finally delivered to the Museum, and Harry and I carted it to Brooklyn, put it in his boat, and motored over to our boat house, arriving at high tide to facilitate unloading. At about the same time a graduate student from the University of Wollongong delivered the constant temperature box to the Brooklyn boat landing, where Harry and I loaded it onto his boat, nearly capsizing it, and installed it in the boat house. Within a day or two of tinkering around, a lab bench was constructed from a wooden door, and everything was assembled for open-flow respirometry, a fancy name for the process we employ for measurement of oxygen consumption. The boathouse at 79 Grantham Crescent had become the Dangar Island Research Laboratory. It wasn't long, however, before the air pump I had brought from the States "packed up" (Australianese for becoming non-functional). This was essential to the system for it supplied air to the respirometry chambers in which birds were to be held for measurement of oxygen consumption, our primary measure of energy expenditure. After two unsuccessful attempts at repair, I resorted to a downtown Sydney hardware store where I purchased a "Little Beaver" air compressor, part of a spray-paint unit. The "Little Beaver" worked like a charm, and we were fully operable.

The field component of the study consisted of two phases: one of locating and collecting experimental samples of birds in mist nets, and the other a series of monthly observations of activity and behavior of Honeyeaters. The first couple of times in the field Harry took us in his boat across the river and around a point to the small town of Patonga, just inside the mouth of the Hawkesbury River. From there, we hiked about 1 1/2 miles to his study sites in the National Park. He showed us places where we might collect birds that would not interfere with his bird-banding and population studies. The habitat there was mixed eucalypt woodland and heathland, the latter being similar to California chaparral. Much of the perennial vegetation in this habitat produced nectar-bearing flowers, particularly the *Eucalyptus* spp. (family Myrtaceae), and the *Banksia* spp. and *Grevillea* spp. (both family Proteaceae). Large numbers of Honeyeaters are attracted to these plants when they are in bloom with their large, showy flowers of yellow, orange, or red; and, in a given habitat, there are almost always some that are blooming. In the process of locating good potential collecting sites, we also found a series of places for our observational work. In the meantime I had applied to the New South Wales National Parks and Wildlife Commission for a scientific collecting permit, allowing us to capture birds for measurement, and then releasing them at the site of capture at completion of measurement. I had specified the number of individuals of each species I wished to collect, followed instructions carefully, and was rewarded not long thereafter with the issuance of a permit. However, on careful inspection I discovered with dismay that the permit allowed me to collect flowers from heathland shrubs and woodland trees, saying nothing about birds. Harry advised me just to carry my flower permit, and a copy of his netting permit to which he would have my name added; "Oh well", I thought.

By then we had the loan of an aluminum skiff with outboard motor for the duration of our residence on the island, but soon found it impractical to use the boat for transport to the study site, even though it only took a half hour to get to Patonga. Instead we found it much more practical to boat to Brooklyn and then drive the "Yeller'roo" on the 1 1/2-hr circuitous trip to the site, allowing us to carry net poles, nets, specimen bags, water, lunch and various gear much more handily. In addition we frequently had to change netting sites during a single day, often a distance of up to a mile. The mist nets we brought from the states for use proved ideal, and were 30 ft. long by 7 ft. high, made from fine, barely visible, black nylon thread with a mesh size of about 3/4 inch. When supported by vertical net poles at either end, and set before or between flowering shrubs or trees, they are very effective at intercepting and entangling birds flying to or from these nectar sources. Careful observation of feeding birds told us where prime netting locations would be, and we were readily able to catch samples of eight species of honeyeater birds that co-occurred, although not continuously, in the same habitat. Some of the species appeared to be year around residents, while others were there more temporarily; during the period of study (February to November 1983), we were able to collect adequate samples of all eight species. Of two additional species, the Noisy Miners (*Manorina melanocephala),* were caught in our windowed verandah on Dangar Island, by luring them in through a window with food and shutting the window behind; the largest species, the Red Wattlebird (*Acanthochaera carunculata*) was netted in the back yard of a museum colleague who lived in north Sydney. The ten species ranged in size from the smallest at 11.4 g to the largest at 126.8 g, over a magnitude in body weight, and therefore ideal for our size-related studies.

In addition to collecting birds, we also had to have a means of maintaining them in captivity during the period of measurement. For this we constructed a rather elaborate aviary at one end of our verandah. It was of 1/2-inch aviary wire mesh and had eight compartments in which birds could be maintained singly, or in groups of three or four per compartment for the smaller species. We had purchased drinking tubes from a nearby pet store to provide the captive birds with artificial nectar, a 20 per cent mixture of table sugar (sucrose) with water. We soon discovered that, while the birds drank this mixture readily, they lost weight and condition rather rapidly on this as the sole food and water source. "Of course," we thought, "they have no protein". So, with a little reading, many phone calls, and some mental meanderings, we stumbled upon the solution or two solutions. A biologist friend Gordon Grigg, then at Sydney University, told me about Amino-lite, an amino-acid rich elixir used to inject into protein-deficient calves and piglets; Gordon had used it on some of his sick crocodiles, with good results. A phone call to a Hornsby veterinarian soon put us onto a good supply of the costly liquid. Then we learned about a mixture of high-protein baby cereal and black currant concentrate, both readily available in the local market. With sugar water laced with Amino-lite, and a slurry of currant pabulum to choose from, our Honeyeaters did very well, maintained or gained weight, and remained healthy throughout the periods of captivity and measurement. After the period of measurement, each was released where captured. This little exercise in captive animal maintenance proved once again that there is always a way, but you just have to find it!

Once we had captive birds and functional equipment, we had to set about the task of making measurements. The respirometry system was set up to measure three birds simultaneously, each in a separate chamber, with a fourth empty chamber to serve as a control. Barbara and I took turns in the laboratory with both morning and afternoon measurements, breaking only for lunch. Barbara did the lion's share of the measuring, and became so adept that she was able to knit a baby blanket for our yet-to-be-born during the process; as her pregnancy continued she finally had to turn sideways to the instruments while measuring. At times one of us would make measurements all day, while the other traveled to the field site to collect or release birds. Occasionally, while netting, we would catch an entire experimental sample (six birds) of one species at the same time in a single net. On one occasion a whole flock of ten White-naped Honeyeaters (*Mellithreptus lunatus*) was caught simultaneously. The inhabitants of Dangar Island were unaccustomed to anything out of the ordinary, leave alone fairly sophisticated research equipment, so it was common for different residents to drop by for an inspection and explanation. After a while, the Dangar Island Research Laboratory became an established (ordinary) part of island life. Even the Director of the Australian Museum came by for an inspection of his Visiting Curator of Ecology, and paraphernalia. We think he was favorably impressed!

Once a month we broke our routine with observational studies at two sites at Brisbane Waters National Park, one a woodland site and the other an open heathland site. These were observations of social behavior and activity during the daylight hours, categorizing the kinds and numbers of birds seen, their interactions and a variety of activities in which they engaged during the daylight hours. These were designed to reveal many aspects of behavior, including the presence (or absence) and intensity of the anticipated dominance hierarchy. Barbara elected to take the woodland site and I the heathland site. Our initial mistake was to attempt the dawn to dusk observations on a single calendar day in February, midsummer. It was a long, hot, grueling day, and we knew there had to be a better way. Thereafter we drove to the site in late morning, made the afternoon observations and then stayed that night at Sydney University's Crommelin Research Station in nearby Pearl Beach. The next morning we were out early for the dawn to noon observations and then drove back to Brooklyn and boated to our island. That was a much better way. Toward the end of Barbara's pregnancy, we hired one of Harry's volunteers from the Museum to fill in for Barbara. As it turned out, we could detect no consistent dominance hierarchy. This was an awful lot of trouble to go to just for some negative evidence, but that revelation was important, and, that's science!

Not every moment of our lives was devoted to the laboratory, fieldwork and housekeeping for Barb and me, and school for the boys, as we did manage to schedule in a few jaunts and exciting events here and there. One of these was an extended three-week trip with the "Yeller'roo" to Darwin, at the Top End of Australia, to attend the annual meetings of the Australian Mammal Society, at which I was presenting a paper. We had completed about one-half of the scheduled lab work and were quite comfortable about the idea of getting off the island for a while for some exploring; in fact, we were quite desperate to do so. We intended to go up the east coast to Queensland, then across central Queensland into the Northern Territory, and north to Darwin. I had a long-standing invitation from Reg and Marg Arthur, former managers of Sandringham Station, to bring my family for a visit to the cattle station currently under their management, Morney Station, west of Windorah, the gill-netting capitol of the outback! I reconfirmed the invitation, and we planned to include a jaunt through the Channel Country of southwestern Queensland, knowing that it seldom rains there during the late spring month of May. Our four-wheel drive "Yeller'roo" should get us through that country with no worries.

As we would be traveling through some remote country, we had to carry camping gear, and jerry cans with water and extra fuel, in addition to clothing, food, and other things necessary for travel. It required several boat trips from the island and a lot of careful packing to get everything loaded into the back of the station wagon and on the roof rack, leaving enough room for two gangly boys in the back seat, and Barbara and me in the front. As we began to pack, it began to rain, but that's usual around Sydney, fall through spring. We finally got underway by mid-afternoon, heading up the east coast to the vicinity of Newcastle, where we holed up in a motel while the rain pelted down. "We were sure to leave it behind as we head north into the subtropics, where summer rains are more common", we thought. But no, the rain just followed us up the coast, through the banana belt in northern New South Wales, and into Queensland. "This is beginning to seem familiar" I thought, as I remembered the rain that had descended upon me nearly every time I came to this dry country. Barbara began to look at me dubiously. Even though it was raining, Chris and Steve wanted to see Queensland's famous Gold Coast and Surfer's Paradise, just south of Brisbane. They had grown up near California's Huntington Beach and Laguna Beach, both decent surfing beaches, and had been led to believe that Surfer's Paradise would live up to its name. Upon inspection, we all agreed that it was okay, but nothing special. In fact, it looked a lot like Huntington Beach with high rises.

We had planned to head west into central Queensland from Brisbane on the Warrego Highway to Charleville, and then on to Windorah and Morney Station along the Diamantina Developmental Road (see map 4). A telephone call from Brisbane to the Queensland Roads Condition recorded number convinced me otherwise, for all roads west of Charleville were flooded, and the road from Brisbane to Charleville was questionable.

Another possibility would be to go further up the coast to Rockhampton, and then inland from there, but that seemed doubtful too, as we would have to cross flooding Cooper Creek, and traverse soggy dirt tracks. At least Rockhampton would put us closer to our destination, Darwin. When we reached Rockhampton, it was still raining, and it was apparent that we were in a far-reaching downpour, covering much of Queensland. There was no point in trying to get into the interior from there, so we stopped briefly to do some last minute birthday shopping, as Barbara's birthday was the next day, and then headed further up the coast. At Rockhampton, we crossed the Fitzroy River, just before it reaches the Pacific Ocean, and it was beginning to flood over the road. We put the "Yeller'roo" into four-wheel drive and crossed, with water nearly up to the floorboards. Shortly after crossing, we heard on the radio that the highway on which we were traveling was closed behind us due to flooding of the Fitzroy River; we must have been one of the last cars to get across. By then, we had entered the tropics, and everything was warm and green, but wet, an improvement over cold and wet. We also had been traveling parallel to the Great Barrier Reef, but the weather had been too foul to contemplate even a visit to a local beach. North of Rockhampton, though, the highway proceeds very close to the shore all the way to Townsville, so we hoped to visit some beaches and the boys were anxious to swim again in warm, tropical water. At that point Barbara and I knew it was futile to think further of visiting the Arthurs at Morney Station, at least until our return from Darwin, a couple of weeks later. By then, perhaps, the rain might have dissipated and we could come in from Mt. Isa, the way Tony Lee and I had entered that country 17 years earlier in our rental car, after flying in from Melbourne.

As it was Barbara's birthday, May 6, we had gone to a seafood shop in Rockhampton where I bought more than a kilogram of King Prawns so that we could have a giant shrimp feed to celebrate. Although still threatening, the weather began to look better as we traveled north, after crossing the Fitzroy. We anticipated camping, but not with enthusiasm, as more rain was forecast, and not much is worse than a wet birthday. Not too far out of Rockhampton we were pleasantly surprised to find a new caravan (trailer) park, with campground and amenities block, the latter being a large central building with communal kitchen, ladies and gents showers and toilets, and laundry. This looked too good to pass by, so we ducked in as evening approached. A query at the office revealed that not only were there tent sites, but there were onsite vans for hire. "What are those?", I asked. I was informed that they are self-contained caravans (house trailers) with everything supplied except linens, with power, completely quipped kitchen, lights, etc. In addition, the nightly rental was under AUS $20.00 (less than US$15.00). This was too good to be true, so we opted to splurge on our sabbatical budget, and treat Mum to a dry birthday. The prawns, which were boiled lightly in salted water, were delicious, especially in the company of a fine bottle of white Aussie wine. Barbara was delighted with her presents, which were few in number but rich with intent. One of these was a large T-shirt emblazoned on the front with "DESIGNED IN CALIFORNIA; ASSEMBLED IN AUSTRALIA", to be worn over the increasingly large lump that was to become Ian.

The next day we headed north toward Mackay and Townsville, with the highway initially passing inland, and then along the coast. By now, we were in the tropics, and many of the older residences were "Queensland houses". These are elevated homes built on cement or brick pillars, with open space between the ground and floorboards. They are designed mainly to prevent termites from consuming the houses, which are built largely of wood; this also elevates the house up into the breezes, and provides a rain-free shelter for the "chooks" (chickens). Periodically we stopped along inviting beaches with the boys anxious to swim, particularly since it appeared that we were finally leaving the rain behind. However, each beach was posted with warning signs that read: "Danger! Sea Wasps and Bluebottles Present. Swimming not Advised". We knew that these were small jellyfish (coelenterates) with spreading, nearly invisible tendrils, covered with stinging cells (nematocysts). Sea wasps, in particular, are dangerous, inflicting extremely painful and sometimes fatal stings. Bluebottles are Portuguese Man-o-wars, less dangerous, but nevertheless producing a very unpleasant series of stings. These signs greatly decreased the boys' desires to swim, and we settled for a walk or run along these beautiful white coral sand beaches and coves. Somewhere around here, while stopped at a petrol station, we saw that the proprietor had avocados for sale, and these were huge, nearly melon-sized fruit. We bought one, about the size and shape of a large eggplant, and it turned out to be delicious. At Mackay, we were directed north to Airlie Beach and Shute Harbor, just inside the Whitsunday Group of Islands. We hoped to arrange a boat trip out to the Barrier Reef for a quick inspection, and were told that this was a good place to do just that. Stopping at an information booth just short of Airlie Beach, the gentleman in charge informed us that he could book us on a full-day trip to the reef the following day, aboard a motor-powered catamaran out of Shute Harbor, that was equipped with snorkeling gear and glass-bottomed boat for viewing the reef. We agreed that this would be ideal, and made the necessary arrangements. We were also directed to a beautiful caravan park with extremely pleasant on-site vans situated in a tropical garden at Airlie Beach, one of which we took for two nights. That evening we had a very pleasant seafood dinner at a local restaurant, further celebration of Barbara's birthday. Our cruise to the Barrier Reef the next day couldn't have been much better for a one-day excursion. It was a clear, beautiful day, and the catamaran cruised out of the harbor and out about 20 miles with astonishing speed. It tied up to a buoy over the reef where we were each outfitted with fins, mask and snorkel. An inflatable boat with outboard motor was put overboard with a crewmember as pilot, who would cruise around to make sure everyone in the water was safe. We were advised that there was a rather strong tidal current, and that we should swim up current, away from the boat, so that we could then drift back to the boat, rather than away from it. Then, without mention of sharks, sea wasps or blue bottles, or other denizens of the deep, we were all urged to jump in, without further advice. We could see the reef about 12 feet below the surface, with other portions sticking above the surface about 100 feet up current. We jumped into the warm, clear water and, as far as one could see, the purple, yellow and bluish corals spread out, some calcareous and some soft corals. There were a few colorful reef fish here and there, and the vastness was overwhelming.

We began working our way up current, as directed, toward an emergent part of the reef, and a floating platform. The current was quite strong and Barbara, who was then four months pregnant, began to tire while swimming belly down, even though she is a strong swimmer. Swimming on her back, however, relieved the burden, and we were soon at the platform, near where we spent the next hour exploring the nooks and crannies of the Great Barrier Reef. Its size was the most impressive thing about it, as it seemed endless. The colorful corals and fishes were also impressive, but not as spectacular as those we had seen a few months earlier at the Tahitian island of Huahine. Eventually we drifted back to the catamaran for lunch, followed by a run or two in the glass-bottomed boat, from which Chris and Steve saw a shark, fortunately after we had emerged from snorkeling. By then, it was early afternoon, and it was time to cruise back to Shute Harbor, the end of another memorable adventure.

The next morning we drove up to Townsville along a beautiful coastline, with warm, clear weather, having finally left the rains behind, and then west into the interior of central Queensland. Our route took us along the Flinders Highway, a good sealed road, through grassy plains to Mt. Isa. Soon we were in cattle country, where properties were separated by fences, requiring numerous grids (cattle-guards) on the road. Some of these were elevated above the usual road level, with a ramp running up to the grid on either side. One of these I took too fast in the Yeller'roo, and we were airborne for a short time, coming down hard on the other side, resulting in a near collapse of our roof rack. The rack was mounted with too many heavy things, including jerry cans with fuel and water. We limped into the next town, Hughenden, and rearranged the load, bending back some of the collapsed supports of the roof rack, and thereafter carried only light, bulky items up top. That evening we camped at a lovely spot just beyond Mt. Isa, up against a hill liberally sprinkled with white-barked Ghost Gums (*Eucalyptus papuana*), nearly shining in stark relief to the dark red rocks of the hill behind. Clumps of Spinifex (*Triodia* sp.) formed an intermittent understory, and Spinifex Pigeons (*Petrophassa plumifera*) ran amongst the clumps, with their topknots bobbing, and looking at us curiously. These birds had brought me to Australia and Mt. Isa initially, back in 1966. Our friends the bushflies were annoyingly plentiful, and we waited impatiently for darkness and their retreat to inactivity for the night. The next morning we were awakened by a bright sun and the incessant buzzing of bushflies congregating outside our tents, waiting to make our lives miserable as soon as we emerged. About the only good thing that can be said of bushflies is that they certainly hasten your departure, as you do not tarry long in their presence if you have an option. Ours was a quick breakfast and pack up, and off in the relatively fly-free Yeller'roo.

As we proceeded west toward the interior, the country was progressively more desert-like in appearance, even though we were well within the geographic tropics, at a latitude of about 20 degrees south. We had come basically due west from tropical Townsville, passing through a gradient of subtropical grassland in central Queensland to progressively more arid country and true deserts around Mt. Isa, with the primary variables being distance from the sea and a flat, featureless terrain. Coming from the northern hemisphere, where deserts are "where they are supposed to be", it is difficult for me to rationalize the presence of deserts in the tropics. From Mt. Isa, we headed northwest along the Barkly Highway, and the southern end of the Barkly Tablelands, much of which is covered by Mitchell Grass (*Astrebla* spp.) prairie, resembling very much portions of the Great Plains in North America. We were leaving the desert and entering the subtropics, and this transition was punctuated with an increase in size of the termite mounds from foot-high hillocks around Mt. Isa, to giant, red globular structures up to 12 feet in height as we entered and proceeded into the Northern Territory. Just north of Tennant Creek, N.T. the Barkly Highway deposited us on the Stuart Highway that runs across the continent from Adelaide in the south to Darwin in the north, and we headed toward Darwin. Our first stop was for fuel and, we supposed, lunch at Renner Springs. We stopped just as a tour bus pulled in and immediately as the tourists departed the bus, they commenced the "Great Australian Wave", waving a hand continually across the face, often frantically, indicating that the bushflies were thick. It was a hot, sticky day and Renner Springs was, as the tour bus driver told us, "The bushfly capital of Australia". We soon discovered he was right, as the initial attempt at making a peanut butter sandwich outside at our lunch stop, resulted in a peanut butter and fly sandwich. We hastily beat a retreat to the Yeller'roo, and Barbara made us all sandwiches in the car as we drove on to the north, leaving a stream of bushflies trailing behind.

The remainder of the drive to Darwin was pleasant, but uneventful, and the country became progressively tropical again. Darwin was a bustling little town, largely new after having been rebuilt twice during the twentieth century: once after bombing by the Japanese in 1942 during World War II, and, again, following the disastrous Cyclone Tracy on Christmas Eve, 1974, that nearly destroyed the town. It is very much a tourist town, with expensive hotels and motels, flash restaurants, and numerous other attractions. We booked into a tidy little caravan park where our onsite van was next to the swimming pool. While I attended the annual meetings of the Australian Mammal Society at the nearby casino, Barbara and the boys took in the tourist attractions, and lazed around the pool. The highlight of the meeting was the barbecue, which featured roast Water Buffalo (*Bubalus bubalis*), a now feral beast in the "Top End" (the northern part of the Northern Territory), introduced from southern Asia. They have become so successfully established in the Top End that it is estimated "that well over half - perhaps as much as 90 percent - of all free-ranging buffaloes in this world live in Australia" (McNight 1976). Not only was the meat delicious, but also we were doing the country and its native animals a favor by consuming this marauder of fragile tropical habitats.

After the meetings, we traveled east into Kakadu National Park in Arnhem Land, where we stayed with our good friends Steve Morton and Faye Alexander and their children, Josh and George, in the small village of Jabiru. Steve, at that time, was a government ecologist with the Alligator Rivers Region Research Institute, assessing the impact of uranium mining on the native vertebrates of that region. Steve and Faye had worked earlier in my laboratory at U.C, Irvine, while Steve was on a postdoctoral study program, after completing his Ph.D. at Melbourne University. Not only did our visit provide an opportunity to catch up with Steve and Faye, but it also gave me a chance to view first-hand the strange ecology of the Top End, and for all of us to see the incredible Aboriginal rock art at Kakadu National Park. The Top End is the land of the wet-dry tropics, where everything is dictated by six months of summer monsoonal rain, followed by six months of winter drought, referred to by the locals as in the wet, or in the dry. During the summer monsoons vast area are inundated as the spreading river systems that network across Arnhem Land flood over their banks, leaving huge tropical marshlands. These same monsoons periodically sweep further south, creating flooding in the Channel Country. When the rains cease and winter drought commences, the rivers recede, the marshlands shrink into swamps and billabongs, and the habitat changes from aquatic to terrestrial, providing an ecologist's holiday or nightmare, depending upon how you look at it. The entire area was, and still is to a lesser degree, inhabited by Aboriginals, who reaped the benefits of tropical biotic productivity and diversity. Much of what they saw, caught and ate is depicted in the amazing array of pictographs painted on the walls and ceilings of numerous rock shelters, particularly of fishes, reptiles, birds and mammals. On many of these rock faces, the paintings in natural pigments are superimposed over each other, and are sufficiently realistic, even though often quite stylized, to enable identification of animals to species. We were amazed to see on one rock wall a painting of what had to have been a Thylacine or Tasmanian Wolf (*Thylacinus cyanocephalus*). These were dog-sized carnivorous marsupials, the last known authenticated capture of which was in Tasmania in 1933; they are now likely extinct. Thylacines were once widespread on the Australian mainland and Papua New Guinea, but appeared to have declined in competition with the Aboriginal-introduced Dingo, becoming extinct everywhere but Tasmania not less than 200 years ago (Strahan 1983). This rock-art depiction of a doglike animal bearing the characteristic Thylacine flank-striping, indicates that the Aboriginals of these ancient times in the Top End knew the animal well.

On one of our last days at Jabiru we had a small U.C., Irvine reunion with a picnic at a billabong, where Steve took us all on a ride on an airboat across swampy water inhabited by Salt-water crocodiles (*Crocodilus porosus*) - we saw one about 14 feet long, as the airboat swept across the shallow water. Present at the picnic were Lynn Day, then of Macquarrie University, and Denny O'Dowd, of Monash University, the Morton-Alexanders and the MacMillens. We had all been together as laboratory technicians (Barb and Faye), graduate students (Lynn and Denny), post-doctoral scholars (Steve Morton), or Faculty (me) at UCI during 1978. As we were driving back from the picnic, a giant Water Buffalo charged across the road and stopped about 50 feet away, to our right. I stopped the car and Barb got out the left side (Aussie cars are right-hand drive) cautiously to photograph it over the hood, as we all watched. Just then, I heard Denny whisper from the rear seat, "Dick, look to the left". There not ten feet behind Barb, was another monster buffalo, standing and staring at us. Denny told Barb to turn around. She quickly got into the car. They can be dangerous, but fortunately, this one just stood there placidly as we photographed it from the car, although it was too close for Barb to photograph with her telephoto lens. And this was even before Crocodile Dundee!

It was time to leave the tropics and head south, toward Sydney and our home and lab on Dangar Island. With still some faint hope of entering the Channel Country for a visit with the Arthurs at Morney Station, we drove south to Tennant Creek, from which we could retrace our route to Mt. Isa and then work our way along dirt tracks to Sandringham and Morney Stations, in southwestern Queensland. At Tennant Creek, we called a recorded road-condition report for Queensland, and learned that the Channel Country of the Georgina and Diamantina Rivers was still flooded, with access unlikely. This news only reinforced an uncomfortable feeling that had followed us all the way since leaving Dangar Island, a feeling that developed after receipt of a letter from Reg Arthur just prior to our departure. Reg, in his soft-spoken way, wrote...."at today's date (27 April 1983) we have a reasonable flood in the Farrer's Channels which are between here and Bedourie due to some good rain we all received over the ANZAC week-end (ca. 25 April). Now these channels could be crossable by 10th or 11th May or they could not - it depends on how much water there is coming down from Connemara & Tonkaroo (stations to the north of Morney). These Farrer's Channels are only 18 miles from here but theres about 18 of them & about 6 miles from 1 side to the other & as there is no bitumen it makes it a bit awkward to get over them when they've run. The actual crossings are A.O.K. being cement drop sections but its all mud between the cement that's the problem."

Taken together, all of this was too reminiscent of my earlier experiences in the Channel Country during flooding, ones that I did not want to repeat with my family and an over-loaded vehicle. So, disappointedly, we adopted our alternative plan to skip the Channel Country altogether, continue south along the Stuart Highway to Alice Springs, and then on through South Australia to Port Augusta, before heading west into New South Wales, across that state, and finally home. The road was sealed as far south as Alice Springs, and, with the exception of small stretches of road in Kakadu National Park, we had been on sealed roads throughout the trip, becoming quite spoiled to smooth sailing at 100+ km/hr (ca. 60 mi/hr) and no dust. At Alice Springs, all of this was to end, as the Stuart Highway south of Alice at that time was unsealed nearly all the way to Port Augusta, a dusty distance of about 1226 km (766 mi).

On our way south from Kakadu we camped in caravan parks, or off the road in the bush, as the occasion allowed. A very pleasant stay was at a caravan park in the country town of Katherine, where we saw Aboriginals perform a coroborree in their colorful body paint, and with animated dancing, mimicking various mythological creatures from their Dreamtimes. Next, we camped off the road on a track through a fence and beyond a cattle grid (i.e., guard). It was a cool evening and we had a campfire. Shortly after dark, a vehicle came along the track and stopped opposite our camp. Knowing that we were likely trespassing on station land, I walked sheepishly over to the Toyota truck, driven by a station stockman and his Aboriginal "offsider" (assistant). I explained who we were, and that we only wanted to camp there for the night. The stockman nodded his approval and asked us only to be careful with the fire, as it was drying out and there was lots of vegetation about. I assured him, with relief, that we would be careful, and thanked him. The next morning we were greeted with the buzzing of a million bushflies awaiting our emergence from the tents. Burning green Spinifex seemed to drive them off somewhat, allowing a quick, smoky breakfast before our departure. At Alice Springs, we stayed at the Camel Farm, which had a small caravan park, located south of Heavitree Gap and the town. In 1983 Alice Springs was just beginning to boom as a tourist attraction, but still had some of the feeling of a country town, with a liberal mix of both Caucasian and Aboriginal people, augmented by the increasingly frequent tour buses laden with people from everywhere.

From Alice Springs, we started our dusty drive on the dirt Stuart Highway through the southern half of Australia, nearly all arid country. I soon was able to introduce Barbara, Chris and Steve to bulldust, which is insidiously fine-grained dust that collects in potholes in the road, and then becomes temporarily airborne as vehicles drive over it. It is about the same consistency as powdered cement before adding water, except lighter in weight, and it remains suspended in the air for a long time, entering all exposed human orifices. We soon learned that the only way to keep it out of the vehicle was to drive with the windows closed and the air-circulation system of the car (we didn't have air-conditioning) set to internal circulation with the fan on. This maintained a slight positive air pressure inside the car, keeping the dust out. But it also resulted in a gradual increase in carbon dioxide accompanied by a decrease in oxygen, an increase in temperature and humidity, which were returned to normal by stopping periodically with all doors open.

Much of the commercial transport in central Australia is by semi-trucks and trailers, usually road trains of three trailers in tandem, and these were numerous between Alice Springs and Adelaide. Road trains churn up an unbelievable amount of bulldust, and as their cabs are both air-conditioned and elevated, it is of no concern to the driver who is listening to tapes by Slim Dusty or Midnight Oil, and charging up the track at 120 km/hr, oblivious to all else. With the first one we passed, I slowed as the Yeller'roo disappeared into a choking cloud of orange-red dust, and visibility was reduced to zero. I slowed still further and blundered blindly ahead for 30 seconds or so before I could vaguely make out the margins of the road, within which luckily I had managed to keep the car. This first experience was enough to convince us of the potential danger of the situation, and it was fortunate that no one else was on the road at the same time, near us. Thereafter, whenever a road train or bus came into sight approaching us, we pulled over to the side of the road and stopped, starting again only after visibility was once again restored (fig. 10).

About 200 km south of Alice Springs we took the Lasseter Highway west to Ayers Rock and Mt. Olga, now more properly known by the Aboriginal names of Uluru and Kata Tjuta, respectively. These gigantic sedimentary monoliths rise out of the desert floor about 50 km from each other. Although close together, they appeared to be geologically unrelated structures, as Ayers Rock is made up of fine-grained, vertically tilted red sediments, while Mt. Olga and its associated monoliths are made up of brown conglomerates, still horizontal in their orientation. Except for these two groups of massive outcrops, no other surface features obscure the horizon as far as one can see. At Ayers Rock, we camped at a campground near the Rock. Chris and Steve found small round gourds called paddy melons, and began playing a game, trying to roll them close to a target without touching it. Several Aboriginal children came around, watched them, and then played with them. One Aboriginal child spied the skull of a Water Buffalo we found at the Top End, which I had attached to the front of our roof rack. The child looked at it quizzically, so I rubbed my stomach and said that it was "good tucker". The child's eyes opened wide, but whether it was with

Figure 10. A road train kicking up bulldust along a dirt highway in central Australia.

belief or disbelief I will never know. About three years previously at this same campground, Lindy Chamberlain's daughter Azaria was purportedly taken from her tent by a Dingo, and killed. Lindy's story was found wanting, and she was tried and found guilty of murder and jailed; a public outcry followed, and a retrial declared her innocent and she was released. The story was later made into a movie starring Meryl Streep entitled "A Cry in the Dark ". In typical outback bad taste, the walls of the Gent's dunny in the campground were embellished with crude graffiti, referring to children and Dingoes, which, with my perverse sense of humor, I found both amusing and distressing.

After Ayers Rock, we continued south into South Australia where, up the track from the hotel roadhouse at Marla, we had our only flat tire of the trip. It was Chris' birthday, and at 14 he took great pride in changing the tire on the dirt track. At Marla, we were told it couldn't be repaired but, since we had two spares, we felt quite safe in continuing on to the opal-mining town of Coober Pedy. There we camped in the dusty, dingy caravan park, stopping the next morning only long enough to go into the market for a few supplies. As we were paying for them, Barbara called my attention to a dark, wooden object in a barrel near the checkout stand. Inspection revealed it to be an intricately carved, smoke-blackened Aboriginal message board, or possibly a wooden tjuringa. Tjuringas are sacred objects said to contain an ancestral soul. The clerk said it was for sale at a very fair price, so we bought it to add to our mounting collection of Aboriginal wooden objects, including didgeridoos, woomeras (spear throwers), boomerangs, and animal figures.

Back on the dusty track, we sped south as fast as we dared and were relieved to finally find the road sealed again before arriving in Port Augusta. There we found a nice caravan park near an estuary, breathed in the clean, moist air coming off the Spencer Gulf and marveled at our recent trek from the Arafura Sea in northern Australia to the Southern Ocean in South Australia, right across the continent some 2720 km (1700 mi). After long showers, we went out to a very good dinner at a local pub to celebrate Chris' birthday. The next morning we had to buy a new tire, as the one that had gone flat was beyond repair, and we still had a lot of bush driving to do before getting back to the New South Wales coast, where we had started weeks earlier. After a late start, we headed east toward Broken Hill, N.S.W., spending the night in a funny little country pub in Yunta, South Australia. I had been coveting one of the large yellow road signs bearing a silhouette of a kangaroo, warning of their probable presence along the highway for the next several to many kilometers. As I had recalled seeing many road-killed kangaroos around Broken Hill, I decided that that was the most likely place to obtain a sign. Being a conscientious, law-abiding citizen, I decided the appropriate way to procure such a prize was through a visit to the local roads department, where they would likely have a discarded one, riddled with bullet holes. Not far from the Yunta Pub was a Department of Roads yard, but no yellow signs. The next stop was with the Superintendent of Roads at Broken Hill, who scratched his head and looked at me peculiarly, and then led me out to a pile of discarded signs, saying "help yourself, mate". There were many yellow signs, lots riddled with bullet holes, but not a single one that was yellow, with bullet holes and a kangaroo. Seeing that I was discouraged, one of the laborers who was standing by said to me "just yank one off the post along the road, like everyone else does, mate!" It was not far out of Broken Hill that we procured the perfect sign, announcing an abundance of kangaroos for the next 170 kms. There it was hanging upside down with only one bolt attaching it to its pole. The last night of the trip was spent at an old two-story, wood-frame pub at Nyngan, New South Wales, a picturesque country town in sheep and wheat country. The next day saw us across the Blue Mountains and out onto the coastal plain of the Hawkesbury River, to Brooklyn. Quite an eventful trip of some 8,934 kms (5,584 mi), that had circumscribed the eastern half of the country, even though the Channel Country within had eluded us.

Another break in the monotony of lab and field work was a two-week seminar junket in August, during which we drove the "Yeller'roo" from Sydney to Canberra to Melbourne to Adelaide, and then returned across the arid zone of central New South Wales, through Broken Hill, Nyngan, Dubbo and back to Sydney. This was a loop that meandered through the southeastern one-fifth of the country, and included the Australian Capital Territory (Canberra), and the states of New South Wales (Sydney), Victoria (Melbourne) and South Australia (Adelaide). On this trip, I presented seminars on my North American rodent work at several institutions along the way. In Canberra one was given at the Commonwealth Scientific and Industrial Research Organization's Division of Ecology and Wildlife (CSIRO), hosted by our good friend Alan Newsome, a wildlife biologist there. This was followed by a visit to Monash University where Tony (Saltbush) Lee, then a Senior Lecturer, arranged for my seminar and our pleasant visit there. It was good to see many of my old friends from 1966-'67, still at Monash. Lastly, I presented a seminar at the Flinders University of South Australia, near Adelaide, arranged and hosted by my former Ph.D. student Russ Baudinette, then a Lecturer in Biology at Flinders; Russ was also an honours graduate in Zoology from Monash, where he did his undergraduate work with Tony Lee. The reader may get the impression that academics is an internationally nepotistic business, and that would not be far from the truth, but purely on an intellectual level. Also, while at Adelaide we caught up with our U.C., Irvine friends and colleagues Al Bennett and Rudi Berkelhamer, there on sabbatical leave at Adelaide University, and accompanied by their daughters. Rudi and Al had arrived in Australia after we had had time to get settled on Dangar Island, and then stayed with us there for a short time before moving on to Adelaide. In fact, Al celebrated his birthday on the island, and attending the party was Bill and Ginny Dawson from the University of Michigan. Bill was on sabbatical leave with Terry Dawson (unrelated) at the University of New South Wales; Bill had been Al Bennett's Ph.D. supervisor at Michigan, and also had been one of George Bartholomew's earlier Ph.D. students. George Bartholomew was also my Ph.D. mentor, so that made Bill my academic brother and Al my academic nephew; the knot of nepotism draws tighter!

Following a successful series of seminars, it was then time to head into the dry country and then back to Sydney. From Adelaide, we drove north and east into New South Wales, passing through the mining town of Broken Hill. Through Terry Dawson I had arranged for us to spend a couple of nights at the University of New South Wales' Arid Zone Research Station at Fowlers Gap, north of Broken Hill, where I had visited on several occasions in 1974-75. On the way to the station, along the still unsealed Silver

City Highway, we spied at least a dozen Shingleback Skinks (also called Stumpy-tails or Sleepy Lizards; *Trachydosaurus rugosus*). We stopped to photograph each of the boys and Barbara holding one or two of the foot-long critters. It must have been then that Ian, still unborn, began to develop his passion for lizards, through maternal impressions. No one, except the cook and the caretaker, was at Fowlers Gap, and the cook, a very friendly lady, was anxious for company. She showed us to our rooms and prepared for us a fine dinner of chops, it being a sheep station. She was somewhat apologetic about foisting lamb upon us until she learned that we loved it and that it is almost a luxury item in the states; with that she promised us roast leg of lamb the next evening, much to our pleasure. I wanted to show Barbara, Chris and Steve a bit of the country, so the following day we drove north along the Silver City Highway to the Packsaddle Pub, about halfway between Broken Hill and Tibooburra, the route along which Mike Augee and I had chased lizards in 1974. Chris had become quite interested in Aboriginal items, and wanted to know whether there might be any old campsites around. I was only aware that Aboriginals had once been throughout that region, and often camped near sand hills that offered, at times, abundant sources of bush tucker (food). I told Chris that we would stop at the first good sand hill, knowing that there were some between Fowlers Gap and Packsaddle, but I silently doubted that there would be any evidence of Aboriginal habitation.

Upon stopping at the first large sand hill we were soon engrossed in chasing lizards, first another Shingleback Skink, and then Chris went bounding after a Bearded Dragon (*Pogona vitticeps*). The dragon stopped under a shrub and Barbara, Steve and I approached to look and photograph it, while Chris went on across the sand hill. Soon we heard Chris shouting, "I've found one, I've found one", obviously very excited. Having dismissed in my mind the possibility of finding anything Aboriginal, we approached, expecting to see another lizard. There he was, crouched over a magnificent Aboriginal grinding stone of flat sandstone about 1 1/2 ft. long and 3/4 ft. wide, with two parallel grooves worn into it from grinding seeds, with a hand held stone pestle. Chris explained that he was chasing another lizard, when he saw this large flat stone on the surface of the sand. Upon turning it over and seeing the worn grooves, he knew immediately what it was, and began to shout. Chris had every reason to be very proud of his find. We then proceeded to the Packsaddle Pub, which had modernized considerably during the nine years since my last visit. When I was last there it was not licensed for drinking on the premises, with only over-the-counter sales technically permitted. However, the publican had an arrangement with a mate in Broken Hill to call him whenever the inspector passed through, so that he could close down the bar shortly before the anticipated arrival. Effectively it was always open for business anyway, so I had to suppose that the inspector seldom appeared. But this time the on-premises license was proudly displayed, and Packsaddle was a bona fide Hotel, for which qualification requires at least one hotel room for hire. The modernization to which I referred was the conversion of part of the building into a hotel room, and the addition of a modern toilet built onto the side, to replace the outback pit dunny behind the building. Barbara and the boys had soft drinks at Packsaddle while, purely for old time sake, I had a Southwark, and then we turned around for our drive back to Fowlers Gap.

That evening we had, as the cook had promised, an excellent lamb roast with potatoes and gravy, with the cook and her caretaker husband joining us for dinner. We were on the road again to Sydney early the next morning, driving pretty much straight through to Dubbo where we stayed in an onsite van in a caravan park. It was next to a drive-in theatre, so the boys watched a silent horror film from the lawn on which our van stood. The next day brought us across the Blue Mountains and down onto the coastal plain to Hornsby and Brooklyn, where we transferred our gear from the car to our boat, and then back to our home on Dangar Island. We had just a little over two months left to complete all of the remaining lab and field work, prepare for and have a baby, and pack for our departure.

By the time of Ian's birth (2 October 1983), essentially all of the metabolic measurements were complete, and the fieldwork was nearing completion. Even though it was apparent that the Honeyeater guild on which we were concentrating was not arranged in any clear-cut social hierarchy, I felt it important to continue the field observations for a full annual cycle. To accomplish this I was able to hire two of Harry Recher's field volunteers to train, and then complete the task after our departure back to the States. Their efforts continued to confirm the absence of any conspicuous social hierarchy based on body size. Our observations of Honeyeater behavior, and flowering intensity and sequence of nectar-producing plants (from large shrubs to large trees), suggested populations of highly mobile birds that covered vast areas of heathland and forest in search of the richest floral rewards. As these were depleted, the birds moved on in search of new rewards. With at least two Families and many Species of trees and shrubs that bore nectar-producing flowers, and with at least some of these blooming at any one time of the year, we concluded that energy in the form of nectar was seldom limiting in supply, and therefore a dominance hierarchy based on size would be unlikely to evolve.

Our laboratory studies, however, proved very interesting and rewarding. Complete measurements of oxygen consumption were made during the daytime (active phase) on samples of ten species of Honeyeaters that ranged in body size from 11 g to 127 g. When these measurements were compared statistically with those predicted for birds of these sizes and their taxonomic group (Order Passeriformes), we found that our Honeyeaters had metabolic rates that were reduced uniformly to about two-thirds of the observed rate for other members of the Passeriformes. Since reduced metabolic rates translate into reduced energy expenditures, these findings were paradoxical and contrary to our field observations that energy in the form of nectar was seldom limiting in the birds' habitats. Why then would the birds seemingly reduce their energy expenses if their needs were so readily met? To get a better handle on this question I then conducted a statistical analysis of the metabolic rates of the major groups of nectar-feeding birds (American Hummingbirds, African Sunbirds, combined with our Australian Honeyeater data). This revealed that all of these nectar-feeding birds had metabolic rates reduced to about the same extent as those found in Honeyeaters, leading to the conclusion that nectar-feeding birds in general have similarly reduced levels of metabolism independent of their taxonomic affiliation, geographic distribution, and energy supply. Again, the reduction of energy utilization in these birds of specialized feeding habits appears unrelated to the amount of energy provided by their environment. I speculate that this reduction is related more to a reduced rate of all bodily functions, resulting in conservation of other limiting but vital substances. Nectar, while rich in sugar-derived energy, is very poor in protein. Nectar-feeding birds rely primarily on insects as protein sources, and insect availability is much more erratic on a day-to-day basis than is nectar, and therefore more likely to be a limiting factor. Reduced rates of energy metabolism would concomitantly result in reduced rates of protein turnover, and so I wonder if protein availability is the answer to my question? Certainly our observation that captive Honeyeaters required protein supplementation of their artificial nectar diet, suggests that protein is as important to their diet as carbohydrates in the form of sugars. Perhaps a reader of this chapter will set out to find the elusive answer to our question of reduced metabolic rates in nectarivorous birds.

Our study of Honeyeaters led to my teaming up with New Zealand ecologist John Craig, then Senior Lecturer at Auckland University, in arranging an international symposium on Honeyeater Ecology at Auckland University in April 1985. This provided an opportunity for another quick visit to New Zealand and Australia. Our Honeyeater work and that of nine others was presented, with enthusiastic responses.

SUGGESTED FURTHER READING:

Craig, J.L., and R.E. MacMillen. 1985. Honeyeater Ecology: an Introduction. New Zealand Journal of Zoology 12: 565-568.

MacMillen, R.E. 1985. Energetic Patterns and Lifestyle in the Meliphagidae. New Zealand Journal of Zoology 12: 623-629.

Chapter 5

IN PURSUIT OF PARROTS, 1989

Crossing the plains we were always aware of a brilliant variety of parrots that flashed in the sunlight. At the windmill and tank of almost every bore we disturbed flocks of Galahs. Those gray and rose-pink cockatoos swirled into the sky, seemingly in thousands, the early sunlight turning their breast feathers to a rich and glowing madder.
Jock Marshall 1966, in "Journey Among Men"

In 1985 I was invited by Russ Baudinette and the Flinders University of South Australia in Adelaide to spend five weeks there in late summer as a visiting research fellow, collaborating with Russ and my long-time research associate Dave Hinds from California State University, Bakersfield, who was there with his family on sabbatical leave. Barbara, Ian and I made the trek from California to Adelaide and back in September and October. At Flinders, we were comfortably housed in a guest flat at the University, where Ian celebrated his second birthday, returning to the country of his birth after two years. We were working on studies of maximal metabolic effort in birds and mammals and, what with my short time and Russ' substantial menagerie at the Uni, we didn't get out bush much, except to pluck Fairy Penguins from their burrows at nearby Victor Harbor and a quick visit by ferry to Kangaroo Island.

We did have one very interesting quasi-official trip to the bush country along the South Australia-New South Wales border, north of Broken Hill, N.S.W. Our very good friend and my long-time mate Alan Newsome, wildlife ecologist with the Commonwealth Scientific and Industrial Research Organization (CSIRO) in Canberra, had been working on the interrelationships between Dingoes, introduced European rabbits, and grazing lands in the arid zone for many years. He had a long-term, ongoing study in which he was monitoring these interrelationships on either side of the Dingo fence along the South Australia-New South Wales border, and had invited us to come up from Adelaide to meet him and his crew at their study site. Alan's study is a difficult, complicated one, but one of critical importance to preserving some kind of ecologically reasonable compromise between native scrublands and their faunas: domestic livestock, feral pests (rabbits), the wild dog, and native animals. Dingoes are looked upon variously as ruthless killers of livestock, effective predators of rabbits, and a naturalized part of the native fauna. The situation is complicated still further by the Dingo fence, with large numbers of Dingoes and few rabbits on the South Australia side, and few Dingoes and large numbers of rabbits on the New South Wales side; additionally, that part of New South Wales tends more to be sheep country, while adjacent South Australia leans toward cattle. Add to this prevailing aridity together with unpredictable, aperiodic rainfall with occasional flooding, and you end up with sufficient variables to constitute an ecologist's nightmare.

Alan Newsome has specialized in vertebrate predator-prey interactions in Australia, and is one of the country's leading experts on this very interesting aspect of population ecology. He, his wife, Jane Thompson, and boys, Andrew and Thomas, had spent six months with us in California while Alan served as Regent's Professor in my Department at U.C., Irvine, giving us a chance to renew our long-standing friendship. There we had discussed the possibility of some bush meanderings together, and my invitation to Flinders provided the perfect opportunity, as it coincided with one of Alan's monitoring trips. Besides, we needed some rabbits for our maximum metabolism studies, providing additional justification for Russ Baudinette and Dave Hinds to make the trip, too. Off we went, with Russ, Barb, Ian and me in a University Subaru station wagon and Dave and Annette Hinds, their son Patrick, and Russ' Ph.D. student, Peter Frappell in the Hinds' Mitsubishi station wagon, bound for Avenal Station in New South Wales. From Adelaide, we traveled northeast toward Broken Hill, N.S.W. on a good sealed road. Just short of the New South Wales border, still in South Australia, we turned north onto a good dirt track for about 130 km, and then east through the dog fence and across into N.S.W. through Brougham's Gate, and then several more kilometers to Avenal Station. Alan greeted us warmly, introduced us to his crew, and showed us around the outbuildings of this large sheep station. Alan and his crew were assigned the shearers quarters, consisting of a series of bunkrooms and a kitchen-dining room. The MacMillens were assigned one bunkroom, the Hinds another, and Russ and Peter a third, all quite snug and a relief from the incessant bushflies outside.

Ian immediately began exploring around and, at age two, was fascinated with everything in nature, and still is. Much to Alan's horror, Ian found and carted around a dead, mummified Galah which, according to Alan, "was certain to have died of parrot fever", but there were no pathological consequences. The following morning Alan showed us around the station a bit, through some red sand hill country where we had a memorable picnic lunch, with Ian and Patrick dodging flies and digging in the red sand, and then back to the station where the manager had presented us all with a magnificent mutton leg of about seven or eight pounds. Barbara and Annette volunteered to roast it in the wood burning range and, after liberally inserting slivers of garlic into it, popped it into the oven, soon filling the kitchen area with the mouth-watering aroma of roast lamb. Annette, Barbara and I kept the range stoked with wood for several hours and then declared the roast done. Barbara and Annette prepared the rest of the meal and rang the dinner bell. It was such a huge leg that we looked forward to leftovers for sandwiches, but clearly underestimated everyone's appetite. We all ate heartily, polishing off the vegetables, potatoes, gravy and roast, leaving nothing but polished bones. After dinner Dave, Russ, Peter and I joined Alan and crew for some rabbiting - spotlighting from a vehicle for rabbits, then getting out and running them down on foot. This is easier said than done, and Dave and I had several misses, none of them near. Peter Caitling was the champion, and Peter Frappell caught at least one, soon giving us a total of five for our studies back at Flinders.

The next day Alan and crew went out to work on their project, while the rest of us elected to explore leisurely around the station and care for the rabbits. We had been advised that the station manager and his wife were going to put on a barbecue for us all that evening, to which we looked forward. Alan told us that it would be next to the shearer's quarters and, since I hadn't seen a barbecue around there, I reckoned it must be a portable one. It was, for in the afternoon the manager arrived with a small tractor, scooped a shallow trench in the sand, with a hummock of sand on either side. Then he maneuvered the tractor blade under a large, thick slab of sheet iron, placing it across the shallow trench where it was supported by the sand hummocks, and informed me that that was the barbecue. It was portable after all, but I had never seen a barbecue assembled by tractor. That evening, after the CSIRO boys arrived back, the manager lit a gum wood fire, heating up the metal plate. Tables and chairs were assembled outside, and beer, wine and soft drinks were copiously available, supplied and enjoyed by all present. Virtually the remainder of the mutton whose leg we had eaten had been rendered into chops, and the manager placed them on the grill. Then I understood why a solid grill rather than a grate was used for such purposes. The chops were well laden with fat, and the grilling process was a combination of cooking the meat and rendering the lard, which flowed off the grill to the side, onto the sand. If a grate had been used, the lard would have produced a veritable conflagration, incinerating the meat. We all sat down to lamb chops and other delectables prepared by the manager's wife. This was obviously an all-you-can-eat feast, and it was delicious. That evening was devoted to yarns, and early to bed for our departure back to Adelaide the next morning. And rather than retracing our way back home exactly, and upon Alan's advice, we headed south from Avenal along a dirt track in New South Wales to Broken Hill, and then by paved road back into South Australia toward Adelaide. On this road at the border there is a quaint little town named Cockburn. It consists of a pub and a small store on the South Australia side, and a petrol station in New South Wales. Cockburn literally sits astride the border; we reckoned that with a name like that you'd want to share your discomforts with both states (it actually is pronounced "Co'burn")!

The time in Adelaide in 1985 passed quickly and productively, and gave us an opportunity to explore a little of Adelaide, a charming city which attracted us very much. It also gave me an opportunity to observe Flinders University, its School of Biological Sciences, and Russ' laboratory and level of organization, all of which I found very impressive.

As we had felt welcome and very much at home in Adelaide and at Flinders University, it was only natural that we arranged to return for my six-month sabbatical leave in 1989. I had recently completed a theoretical paper on a model of water economy in seed-eating birds (MacMillen 1990), and I proposed to Russ that we test the model on Australian seed-eating parrots. The seed-eating parrots of Australia are very diverse, ranging in size many fold, and are of very ancient ancestry, probably traceable back to the time Australia was part of the great southern continent Gondwana (Christidis, et al. 1991). Since the model was based on empirical data derived largely from North American seed-eating finches and sparrows, it seemed to us that an ideal independent test of its validity would be with a diverse group of only remotely related seed-eaters from another continent. Australian parrots should fit this bill admirably. Secondly, I knew that I could step into a fully equipped lab, and would not have to rely upon the rather makeshift, but functional, arrangement we had fabricated for the Honeyeater studies at the Dangar Island Research Laboratory (or boathouse!). Thirdly, after repeated excursions to Australia, and with a much more thorough familiarity with its fauna and geography, I felt confident that I could rely more on a designed-in-advance study, and less on serendipity, although the latter was bound to play some role.

After a brief stopover in North Island, New Zealand, to visit with our good friends Ann Stewart and John Craig and to tame some rainbow trout in Lake Taupo, we ventured on to Adelaide. I had promised Barbara and

Ian a much more normal existence than the one we had experienced on Dangar Island, and Russ had arranged for us to move fairly directly into a temporary rental house for the first several weeks, to be followed by a more permanent rental for the rest of our stay. The first of these was an ordinary but comfortable house in a suburb situated between the University and the sea (Gulf of St. Vincent in the Southern Ocean). The second was a delightful home in the suburb of Blackwood in the Adelaide Hills above the University. It belonged to a Geography Professor at Flinders University who, coincidentally, was on sabbatical leave with his wife in California. It was nicely and comfortably furnished with a large yard full of lizards, much to Ian's delight, and a fireplace for those cool Adelaide nights during fall and winter.

We spent the first several days and nights with Russ, his wife Kaye and their two-year old daughter Emma, giving us a chance to reorient ourselves to the Adelaide area. We also bought a 1985 Nissan dual-cab pickup with a camper shell (in Australian, a dual-cab ute with canopy); as it was silver in color, it promptly became the "Silver Bullet". It would serve us well as a five-passenger vehicle, plus lots of storage space in the rear bed and shell. All of the major Japanese automobile manufacturers export dual cab pickups to Australia, where they are extremely popular. It is a mystery to us why it has taken so long to market them in the United States; they finally arrived in 2000, and are becoming increasingly popular. Our arrival was during the summer school holidays and we moved to our home in the Adelaide Hills about the same time school started. Ian was five and had been enrolled in pre-school up to the time of our departure from California, and it was time for him to start public school. He was enrolled in reception, South Australia's equivalent of our kindergarten, in Trish Pratt's class at Blackwood Junior Primary. The school had only grades reception through two, and each class was a mixed-grade class with all three levels. Each child was urged to develop to his/her fullest capacity and, under Trish's tutelage, Ian flourished, and at the time of our departure 4 1/2 months later he was on the verge of reading and writing, had mastered the decimal monetary system, addition, subtraction and simple fractions, and was inventing multiplication.

As we were only going to be in Australia for six months, during which we hoped to collect at least five species of parrots and employ them in experimental work, I knew that we would have to get cracking early. Russ was a master of organization and soon had me in touch with the right person at the Office of South Australia National Parks and Wildlife for issuance of the necessary scientific collecting and holding permits. This time I was issued a collecting permit for birds, rather than for plants as had happened earlier when working with Honeyeaters. We were also advised that the Cooper Creek area around Innamincka, at the end of the Strzelecki Track, might be a good place to collect birds, as several species of seed-eating parrots were plentiful there. Waterholes and billabongs abounded there, and setting mist nets on the edges of bodies of water in the dry country was the best way to catch birds, as they flew in to drink. Mist nets are about 7 ft. tall and 18 to 30 feet long, woven from fine strands of synthetic fiber, with mesh-size geared to the size of the desired bird species. Even when the net color is black it is nearly invisible when set against a backdrop of water, soil and/or vegetation, and held in place by vertical mounting poles. When set in birds' flight paths it is hoped that they will fly into the net, becoming entangled until removed. Mist-nets can be very effective and are generally harmless. Although our departure for the Cooper would be held up until the arrival of my equipment from California, we made plans for a late January departure. Included in the equipment were some of our camping gear and the mist nets, both essential to the trip. A call to the shipping company indicated that the equipment had arrived in Sydney some time before, and was on its way to Adelaide, "no worries". "By Camel train", I thought. At any rate, it finally arrived and was extricated from customs. Also included in the shipment, of course, were several cans of Mexican salsa, vital to our survival in Australia! With collecting permits and mist nets in hand, we were ready to tackle the Strzelecki Track.

Numerous infamous outback tracks traverse central Australia in many directions and to many destinations. In South Australia the most notable of these are the Oodnadatta Track and the Birdsville Track, to the west and east of Lake Eyre, respectively, and the Strzelecki Track to Innamincka and Cooper Creek. These were originally cattle droving routes, along which cattle were driven by horseback to railheads or pickup points for transport to the markets in Adelaide. Now that huge semi-truck road trains have replaced cattle droving, these tracks are used for trucking, and are increasingly popular challenges to the four-wheel drive crowd of adventurers. The Strzelecki is particularly popular as its southern terminus is nearest Adelaide, it is the shortest of the three, it ends at Cooper Creek and Burke and Wills country; and at its nominal end, Innamincka, there is a pub with meals, accommodation and fuel. Innamincka is only a few miles downstream on the Cooper from the separate sites where Robert O'Hara Burke and his second-in-command, William John Wills, perished within a few days of each other in June 1861 from malnutrition. For us it represented a fairly easy-to-reach arid zone habitat with waterholes, where we were likely to encounter at least three of the more common seed-eating parrots: the small Budgerigar (*Melopsittacus undulatus*), the intermediate-sized Cockatiel (*Nymphicus hollandicus*), the larger Galah (*Cacatua rose-*

icapilla), and perhaps some others.

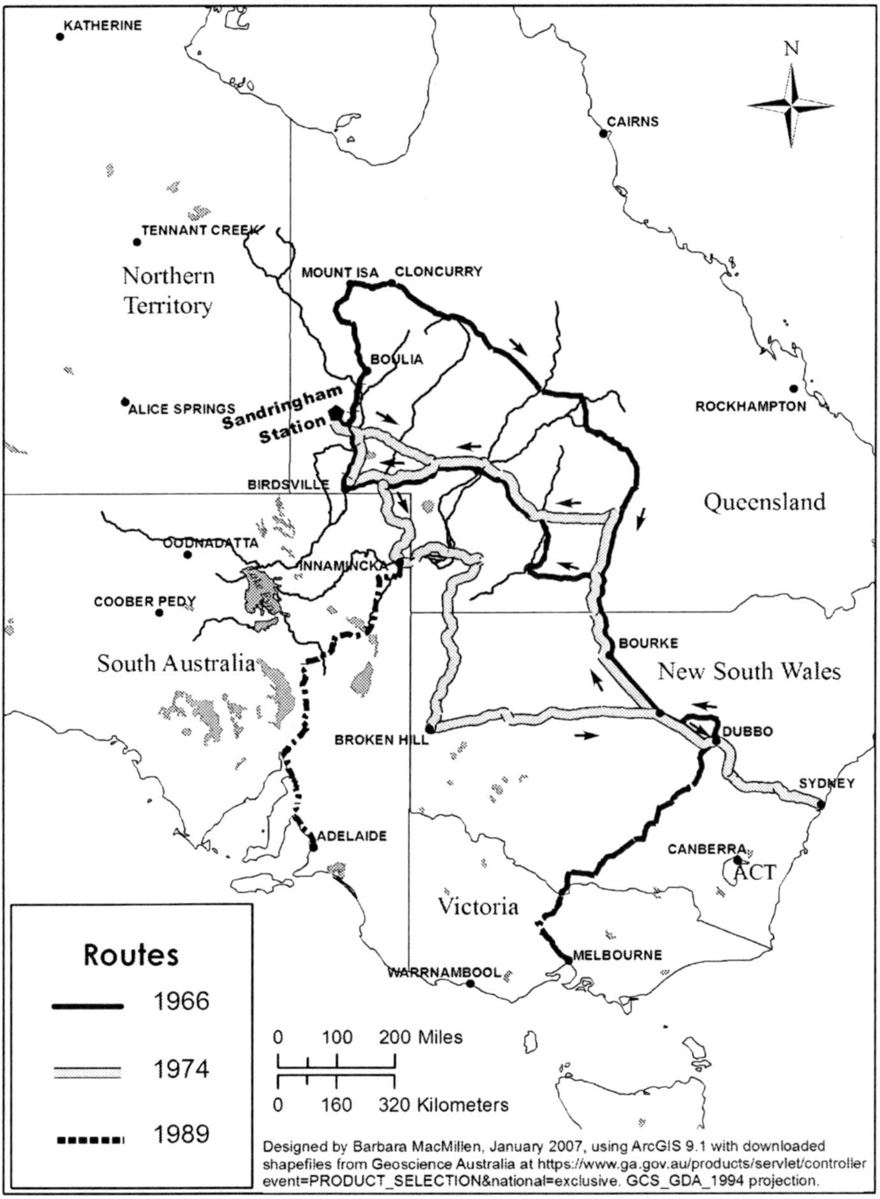

Map 5. The route of our 1989 collecting expedition from Adelaide to Innamincka and return. Also shown are the routes of our earlier expeditions.

To reach the Strzelecki, we traveled north from Adelaide in a University diesel Toyota Landcruiser, nearly to Port Augusta, then Quorn, Leigh Creek and Lyndhurst, all on sealed roads (see map 5). Participants in this expedition were Russ, Barb, Ian and me, with camping gear, food and supplies for a week, mist nets, support poles, and caging for the birds. About three hours out of Adelaide we stopped for a pee, and Barb asked me the whereabouts of our two new, $30.00 each sleeping pads that I had carefully strapped to the roof rack. The straps were there but the pads were not, and it was obvious that they had loosened and blown off. There was nothing I could do to salvage that error, but fortunately, we had a less costly piece of foam for Ian, and persuaded him to share it with his mom and dad, all sleeping on it with at least the upper parts of our bodies. After fueling up at Lyndhurst we turned up the Strzelecki, which passes first across rocky undulating country forming the northern flanks of the Flinders Ranges, and then into more typical desert country. The track was well graded but dusty, and, as evening approached, we came to a large dry creek with stately River Red Gums where we decided to stop to camp. As we sat up camp Ian immediately began to explore, bringing in sheep skulls, cow bones, and the proximal part of a kangaroo's tail skeleton - he was in his kind of place! Our cheap Aussie tent served us well that night, and the sparse foam pad wasn't too uncomfortable. We were up and away early the next morning, eager to find a place to net birds. As we proceeded out onto the desert, passing between Lake Callabonna and Lake Blanche into which Strzelecki Creek empties, both largely dry then, we entered mixed sand hill, clay pan, country, with occasional waterholes. These supported fairly large populations of aquatic birds, but not many parrots, probably because of the near lack of trees. Most parrots require trees for roosting and tree holes for nesting, and prefer areas where there are both trees and waterholes. Soon, though, the track turned north toward Innamincka, along Strzelecki Creek. The creek, although dry, was lined with numerous trees, both gums (*Eucalyptus* spp.) and wattles (*Acacia* spp.). We began to encounter the occasional waterhole, too, and parrots. Galahs, in increasingly large numbers, were present, along with occasional small flocks of Budgerigars and solitary Little Corellas (*Cacatua sanguinea*), and lots of crows (*Corvus* sp.). As we were just about at the halfway point in our fuel supply, we decided to stop at the next attractive-looking waterhole to camp, and to try our luck at netting. If successful, we could then turn back with our catch toward Adelaide; otherwise, if we went any further, we would be committed to going the additional 200 km to Innamincka, which would then be the nearest source of fuel. Soon we came to a very attractive waterhole with a scattering of gums and wattles nearby, and Galahs that flew up from its banks as we approached. This looked ideal, so we headed for a nearby clump of wattle trees for a shaded campsite, got out our gear, and optimistically began setting four nets around the edges of the waterhole. This gave Ian a chance to play in the mud and water for a while before he was banished back to the campsite so as not to frighten birds coming in to drink. The Galahs that we frightened from the banks now sat high in the nearby gum trees, eyeing us suspiciously. This was going to be a piece of cake, I thought to myself.

Once the nets were up we got our camp organized, with bottled gas stove and lantern, folding table and camp chairs, and our tent with the sparse foam pad. Ian played in the shade of the wattle trees with toy cowboys and Indians he had brought for the occasion. We kept an eye on the nets, expecting any moment for a few Galahs to come swooping in to drink. "Here they come," said Barb, and we watched them fly in, between two of the nets, have a leisurely drink, and then fly away. "Oh well, there are plenty more where they came from" we thought, our hopes still high. Then another batch came in, flying directly toward a net, just with the right trajectory to slam into it; this was it. About 10 feet from the net, though, they veered away to the side, alighting next to the pond away from the net. They had obviously seen it, and had avoided it. These birds, though, were not in such a hurry to leave, and padded around in the mud near the water's edge, with their awkward parrot gait, until a couple were actually between the net and the water. This net was on the side of the pond opposite to our camp, and we thought that if we ran to the bank, yelling, we could frighten them into the net. After giving the Galahs a few more minutes and alerting Ian to our plan, the birds in front of the net were joined by a couple more. We rose up in unison and ran shouting to the edge of the pond, Ian thinking this was great fun. The Galahs were indeed startled, and they rose up as one, vertically, instead of angling into the net as we had envisioned, and then merely flew off squawking to a nearby tree. These were damn smart birds!

We decided to reset the nets, paying more attention to their visibility with the muddy water as background, and giving Ian another chance to get wet and muddy, his favorite pastime at age five on a hot afternoon. This time we walked behind each net, insuring that it was as inconspicuous as we could make it, stopping frequently to rescue Ian's wading shoes, slip-on canvas shoes that tended to remain behind embedded in the sticky wet mud as he marched on. Once we were sure of the best set possible, we returned to camp to prepare dinner, and wait for the evening flight to drinking water, which was certain to occur (Fisher, et al. 1972). As we ate, a few more Galahs came in, evading the nets, but not the massive numbers we had anticipated. By the time it was dark nothing had come near the net except a single Masked Lapwing (*Vanellus miles*), a handsome aquatic bird that stumbled into a net. We decided to leave the nets up overnight to avoid resetting them at dawn, and to make certain we caught the early morning flight to water. We awoke early to the raucous cries of Galahs, thinking, "Now we've got them". After the noise settled down, we got up to view the nets, flushing a few Galahs from the edge of the pond, and noting that there were two things struggling in the mesh. I didn't delay to put on shoes, but ran over eagerly with bare feet to secure our first two parrots, which turned out to be a bat of unknown identity and a Tawny Frogmouth (*Podargus strigoides*). The latter is a strange relative of the Poorwill that forages for invertebrates and small vertebrates at night; it is about the size of an American Robin, and its bill opens with an enormous gape. There also was a very large tear in one of the nets, where something large might have gone right through. Or, more likely, an owl had removed some smaller prey item caught in the net. With that disappointment, we decided to leave the nets up through breakfast, to no avail, and then go on up to Innamincka and Cooper Creek to try our luck.

We pulled into the Innamincka Hotel and stores about noon, having passed along most of the length of Strzelecki Creek, which had a few pools of water, and numerous other waterholes adjacent to the main dry creek channels. The creek has its own smaller watershed, but it also receives massive amounts of water during flooding of Cooper Creek, when the latter overflows its banks, leaving numerous waterholes in its wake. We also had seen progressively more parrots as we got nearer to Innamincka, and our hopes were beginning to rise again. While filling up with diesel, beer, and soft drinks, we were directed to a spot along Cooper Creek with lots of River Red Gums, good camping places, and "mobs of parrots". We drove down to the creek about three km from the Hotel, to a truly lovely spot along the creek - really a river about 30 meters wide, and slowly flowing. The reddish water contrasted with the closely cropped (by rabbits as it turned out) grass that formed a lawn extending 20-30 m from the water's edge, and with magnificent huge gum trees lining the bank. White Little Corellas were everywhere in the trees, Galahs flew by occasionally, and Budgerigars were in moderate abundance. This place looked great as all three species were on our list of potential experimental subjects: Budgies (ca. 30 g) were on the small end of our size-related spectrum, Little Corellas (ca. 400 g) on the large end, and Galahs in between. We were the only ones there, so had our choice of campsites.

We elected to set up camp away from the water, up a small, sandy dry creek bed, so that our activities wouldn't disturb birds near the water. After lunch, we set up nets both near the water's edge of Cooper Creek and along a grassy bank under gum trees slightly above the water's edge. At this latter spot Little Corellas were cavorting about in the trees, occasionally flying in low from the water at net height. This place looked much better, and by evening we had netted several small Diamond Doves (*Geopelia cuneata*), six or eight Zebra Finches (*Poephila guttata*), a single Budgerigar, and three or four smallish parrots we had not seen before, the Red-rumped Parrot (*Psephotus haematonotus*, ca. 55 g.). The Diamond Doves and Zebra Finches were released, but we eagerly kept the Budgerigar and the Red-rumps; all of these species were seed-eaters, and the Budgies and Red-rumps could be readily tested with our model. All of these birds were caught in the two nets set close to the water's edge, with nothing yet in the nets further up the bank. By late afternoon, in the nets at the water's edge, we had caught several more Diamond Doves and Zebra Finches, and four or five more Red-rumped Parrots. We had been about to move the nets on the bank above the water when we caught a Little Corella. All of us were beginning to feel a little more relaxed, as by then we had a Budgerigar, about eight Red-rumped Parrots, and a Little Corella.

When birds are caught in mist nets, some individuals and some species remain completely still, do not become badly entangled, and are removed easily. Others struggle continually or intermittently, gathering netting up in their feet, and becoming entangled to varying degrees. The Red-rumps proved to be past masters of entanglement. Moreover, their size, like that of a robin, was too small for Russell and me to manage with a glove, yet large enough to have a menacing beak. Barbara is much more dexterous than either of us and patient, too; she would hold a parrot with a gloved hand, to which it would attach with its beak, and then carefully, systematically and quietly remove the bird from its entanglement. On the other hand, Russ and I would sit there in the mud, cursing at the bird as it nipped our bare hand (usually at the quick of the nail), attempting to disentangle it from its wad of netting. The trick in removing birds from mist nets is to determine the direction from which they entered, and then deftly back them out. That wasn't always easy with Red-rumps, as some entered from one direction, and others from the other direction, and at least one (I swear!) entered from both directions, simultaneously. On one occasion when we had nearly a dozen Red-rumps in the nets at one time, Russell and I resorted to clipping some strands of the nets with scissors, to get around the worst tangles and to relieve our bleeding fingers, frustrations, and cursing, while Barb calmly worked on, smiling. Ian, playing happily all the time in the water, was quite amused by it all, too. It wasn't long, however, before we had all of the birds removed, thanks to Barb's success rate of ten birds to our one.

This was cause to celebrate, so we furled up the nets, and drove back to the Hotel to treat ourselves to dinner. Our dinner was served in a semi-enclosed, roofed patio next to the Hotel, where we had grilled rump steak, chips (french fries) and salad, along with a fine bottle of South Australian wine. While we were eating we heard a strange clip-clopping sound coming down the track from the north, and looked up to see a sort of covered wagon, drawn by two Camels, and carrying a grizzled old character, approaching the pub. Even though by then I had become accustomed to seeing practically anything in central Australia, we had to get up to inspect this oddity and take some photos. The driver was coming south from around Betoota, and had stopped to get some supplies before moving on. He had a crate full of hens ("Chooks" in Australian) under the rear of the wagon, probably for fresh eggs. Ian was particularly intrigued by the Camels. After a brief look at this curiosity, we headed back to finish dinner before it got cold. It wasn't long before the Camel-drawn wagon went clip-clopping down the track. We then went back to camp as dusk descended, to unfurl the nets for morning and to sit around a campfire for a while before retiring to our tents. Upon arrival at camp, we were surprised to find a neighbor, our friend with the Camels and wagon. He was just returning to his camp, on foot, with a rabbit he had shot, which he proceeded to feed to his two dogs. He returned our wave, but otherwise stayed to himself, with two Camels, two dogs and several hens as traveling companions. One of us was heard to comment, "Only in Australia!" while another responded, "She's a big country, mate!"

The next morning Ian awoke early and Barb decided get up and take him for a walk over to check the nets. Upon approaching the nets, she saw that they were full of birds and sent Ian running back to camp to get Russ and me. We dragged ourselves out of our bags, pulled on our clothes, and ran to the nets. Russ headed for the nets near the water which held a few birds, and Barb beckoned Ian and me to the nets on the bank into which a whole flock of a dozen or more Red-rumps had flown. They were flopping about, kneading with their feet to make the tangles more hopeless. I could tell we were in for it, but this would certainly give us a very good sample of Red-rumped Parrots, which fit our needs perfectly, although we had not planned on them (serendipity again!). In addition, our nets held another Little Corella. Dying for a morning cup of coffee, we set to work at the nets, extricating parrots, while Ian played in the mud and water at the edge of the creek, asking periodically about breakfast. As soon as the birds were removed from one net, we furled it to prevent additional captures, and then moved on to the next. It was getting hot and we needed to get the birds removed from the nets before they became heat stressed. In addition, the bushflies were thick, I was sweating, and we were all hungry. I said, more to myself than anyone, "This is fun?" Barbara looked over at me and grinned, and we both agreed that it sure was. In about an hour and a half we had removed all of our birds, collapsed the nets, and sat down to breakfast, leaving the birds in cloth bags in the shade. Russ' nets had some additional Red-rumps and some more Zebra Finches. Over breakfast, we decided to keep 18 Red-rumped Parrots, the 2 Little Corellas, and the Budgerigar for transport back to Adelaide, releasing the remaining Red-rumps, the finches and doves. Since we didn't know how well the birds would travel, we decided to pack up and leave immediately, while the birds were in optimal condition, and drive straight through to Adelaide, stopping only for fuel, refreshments and toilet.

It was a long haul down the Strzelecki Track, and then on down to Adelaide. We made it in about 14 hours with only one unscheduled stop, luckily on pavement - that was to change a rear tire whose tread was peeling off in the middle, making a strange slapping sound that we couldn't track down. Barbara finally figured out what it must be and persuaded us to stop. Sure enough, she was right, and we thankfully changed it before it went flat. We arrived back in Adelaide around midnight, and had prepared caging at the University in advance so it was an easy matter to leave off the birds, and then on to our respective homes and beds, dead tired. All of the parrots made the trip in good health. Ian proved to be a real trooper on this trip, which could have been formidable for many five-year olds. What a guy! Not long after our return to Adelaide, central Australia began to experience periodic, torrential rainfall that continued through much of February and March. We learned that Cooper Creek flooded, with the Innamincka Hotel perched on a small island in a vast lake that extended down much of the Strzelecki Track. Most of the region we visited was under water for at least a couple of weeks, accessible only by boat. The timing of our trip had been most fortuitous.

Since our collecting permit had not included Red-rumped Parrots, we had to call National Parks and Wildlife to have it amended. This was readily done, and in the process we were offered the loan of eight wild-caught Mallee Ringneck Parrots (*Platycercus zonarius*, ca. 125 g.) from the interior of South Australia, and confiscated from illegal bird-traders. These we accepted enthusiastically, as they were one step larger in size than the Red-rumps. In addition we were advised that one of their biologists, Barb St. John, was studying populations of Little Corellas near Leigh Creek, just south of the start of the Strzelecki Track, and that she might be willing to capture some for us. Another phone call arranged that, and there we were with complete samples of three of the five or six species we needed, and more than ample to commence our experimental work while figuring out how to get the others. The lab work was begun in February, early enough to ensure periodic meanderings to alleviate the drudgery.

Our first diversion was a several day retreat over Easter by the MacMillens and Baudinettes to Kangaroo Island, reached by car ferry about an hour out of Cape Jervis, south of Adelaide. There, Russ had arranged the loan of a quaint, very rustic "shack" (weekend cottage in Australian) from a colleague. It was a pleasant breather, and we had a relaxing, fun time exploring beaches, seal colonies, and sandstone caves. Then back for more measurements. I had known for some time, both from the literature (e.g., Aschoff and Pohl 1970) and my own work that body temperature and metabolic rate in birds are lower at night than during the daytime. The study that Russ and I were conducting required simultaneous measurement of oxygen consumption (metabolic rate) and evaporative water loss (from the lungs and skin), and I had presumed that both would be reduced by the same proportion at night as compared to daytime measurements. That being the case, the ratio between these two rate values would remain the same day or night. It was this ratio that would provide us with the information we sought: in a water balance budget for seed-eating birds, metabolic rate is a predictor of the amount of water produced by metabolism, and evaporative water loss is the major avenue of water loss. In order to avoid a whole tiring series of nighttime measurements, we would have to confirm that the ratio remained the same, allowing for only daytime measurements. A trial run with Little Corellas proved my assumption incorrect; the ratios were literally as different as night and day, and we wanted to account for water gains and losses over the twenty-four hour period. Therefore, we were stuck with full measurements during both day and night, for all species. This added a completely new dimension to our work schedules, hopefully compensated by what to us would be very interesting results.

By late March, we had completed measurements of the Little Corellas and Mallee Ringnecks, and were about halfway through with the Redrumps. The MacMillens declared a vacation and headed for Tasmania for ten days in the "Silver Bullet", via Melbourne and then the ferry to Tassie. Our crossing of Bass Strait to Tasmania was in some of the roughest seas the ferry had encountered in many years. Barb and I were tossed around in our bunks sleeplessly, all night, while Ian slept peacefully through it all. The next morning, with bleary eyes, we followed Ian's bounding steps up to the galley deck for breakfast. While waiting in line at the cafeteria, were advised over the loudspeaker that before entering Tasmania, we would all have to pay an entry tax. We knew that Tasmania considered itself, and was considered by others, to be different from the rest of Australia, but, still, it was a part of the Commonwealth, and a bona fide Australian state. After the second call to pay entry tax, having had a cup of coffee, with passports in hand, I dutifully but indignantly went downstairs to the bursar's window, and queued up (got in line). Quite a few of us stood there, getting more impatient as time passed, with nothing happening. We were told that someone would be there soon to take our money, and issue an entry certificate. "First rough seas, then a sleepless night, only to be followed by this nonsense", I said to myself. Still no one came to the window, but at the same time the loudspeaker began to crackle again. I could sense another complication arising, when over the loudspeaker came a cheery female voice that said, "Prime Minister Bob Hawke has just declared that Tasmania is part of the Commonwealth after all, and therefore you will not have to pay an entry tax to disembark". Then she added, equally cheerily "Oh, by the way, Happy April First!" We had forgotten it was April Fool's Day, and my muddled mind was just barely able to see a little humor in the entire proceeding. I rejoined Barbara and Ian in the cafeteria, where they were chuckling over the announcement, and finished my breakfast. We had a lovely tour of Tasmania, staying at Deirdre Frappell's (you met her husband, Peter Frappell, earlier) parents' beach shack at Swansea, on the east side, for a couple of days, then on to Hobart and then inland to the central lakes district where we stayed at Bronte Park Highland Village. From there to Deloraine, then Devonport to meet the ferry for an uneventful (this time) return to Melbourne, and then back to Adelaide. At Bronte Park, we entertained some enormous Brown Trout (in Bronte Lagoon), but the closest we got was a single, missed strike on a nymph. We returned to Adelaide along the coast and at a small fishing village we bought an enormous live crayfish (Spiny Lobster), which we prepared upon arrival back home, eating it with great gusto for two meals. Following that happy interlude, more work was in order.

During our absence, Russ had continued to make measurements while, at the same time, meeting his lecture schedule and many other academic commitments, and had completed the work on the Mallee Ringnecks. These are truly beautiful birds, about the size of North American Mourning Doves, with bright green bodies, blue backs, an orange breast bar, long green tails with outer blue feathers, green head with a red patch on the forehead, and a pronounced yellow collar around the back of the neck. We were ready to start on the Red-rumps and it was my turn for a concerted effort, particularly since we were scheduled to depart on another trip in about two weeks, this time to attend the annual meetings of the Australian Mammal Society in Alice Springs. I was determined to complete all of the daytime and nighttime measurements on fully hydrated Red-rumps before our departure, so I started cracking and, with Russell's help at night, we managed to complete the job on schedule. Russ was not going to attend the Alice Springs meetings, but instead would stay in Adelaide, gradually reducing the daily water consumption of the Red-rumps until their minimal water requirement was determined. Birds in captivity usually drink far more water than they require, probably out of boredom. We wanted to repeat our measurements on birds drinking a minimal water ration, as we suspected this might result in an altered physiological state. By the time of our return to Adelaide, two to three weeks later, I could make the repeat measurements on the water-restricted Red-rumps.

Another, and probably more compelling, reason to go to Alice Springs was that the trip would take us through vast stretches of desert country, right to the center of the continent, where arid-zone parrots abound. We needed to fill out our experimental sample of, Budgerigars and, hopefully, Galahs, and I was intent upon obtaining a sample of Bourke's Parrots (*Neophema bourkii*, ca. 40 g.) that are supposed to occur in a belt on either side of the border that separates South Australia from the Northern Territory. Our route would take us right through this belt. Both Budgies (Budgerigars) and Bourke's were critical to our size-related survey of parrot water economy because they are the smallest and next smallest, respectively, of the seed-eating parrots. Earlier studies (Cade and Dybas 1962; Lindgren, 1973) indicated that they required little or no drinking water under normal laboratory conditions, suggesting that they are unusually economical in meeting their water needs. We wanted the flexibility of being able to collect parrots in both South Australia and the Northern Territory so additional permitting was necessary. A few phone calls resulted in a Northern Territory permit issued by the Northern Territory Conservation Commission, and an import permit issued by South Australian National Parks and Wildlife, allowing us to bring birds caught in the Northern Territory into South Australia for study. Both of these agencies were extremely cooperative and encouraging, largely due to their appreciation of the very fine work that Russell was producing on the functional attributes of their fauna, and their commitment to furthering research on it.

Our departure was to be on 20 April 1989 and somehow Barbara, in her very efficient, caring manner, managed to arrange on the evening before a super celebration of my birthday at our house, replete with Mexican dinner and chocolate cake. In addition to Barb and Ian, the Baudinettes came, as well as Alan Newsome from Canberra, and Lynn Day from Perth, my former Ph.D. student; they were both passing through on their way to the Mammal meetings. A rousing good time was had by all, and I felt more than prepared to face the next year. The "Silver Bullet" (our dual-cab pickup) was packed to the gills once again, with camping and bird collecting gear, and, after dropping Lynn off at the train station and Alan off at a CSIRO office in Adelaide, we headed for Alice Springs. Our route this time would be largely north through Port Augusta and then into the desert via Woomera, Coober Pedy, Marla, all in South Australia, and then into the Northern Territory via Kulgera (just north of the border), Erldunda (turnoff to Ayers Rock or, more properly, Uluru) and Alice Springs. It was a two-day trip with the first night spent at Coober Pedy in, much to Ian's delight, an underground motel. Coober Pedy is an opal-mining town located in a desolate, nearly treeless area atop some of the richest opal fields in the world. To cope with the nearly intolerable summer heat and aridity, much of the populace lives underground in homes and hovels dug into the soft, opal-bearing siltstone. Following this lead, several of the tourist attractions are also underground, including our motel, which was quite elegant, and the fine, underground restaurant where we had dinner. Among his other naturalistic pursuits, Ian is a rock hound, and was intent upon finding an opal. In the morning as we finished breakfast at the underground motel Barb and I urged him to go out near to where the "Silver Bullet" was parked and look for opals. This was meant more as a diversion for him while we packed, rather than serious fossicking. Out he went, and as we emerged with our bags, he asked, "Is this one?" He held a fist-sized rock that sure enough, had a couple of veins of blue opal running through it. Barb and I were astonished, and Ian was quite pleased in finding what he knew he could find all along. He is the sort of boy who could find the proverbial pony under the pile of horse manure! Later, just before our departure back to the States, I tried to convince Ian that the opal-bearing rock he had found was too large to ship home in its entirety, and that we should break off the part with the opal, to keep. After he reluctantly agreed, I whacked the rock with a hammer, breaking it in half, and revealed a much nicer, nearly gem-grade vein of blue opal right in the middle, and now in both halves. Needless to say, we kept both, and I had to acknowledge that Ian did indeed find the pony.

We had made excellent time on our trip so far. When we had last traveled this route in 1983 (from Alice Springs to Port Augusta), the road was a dusty, graded dirt track most of the way (see description in Chapter 4), but it was now an excellent, two-lane sealed highway. Thank goodness it was, for the earlier rains of February and March that had flooded Cooper Creek, and filled Lake Eyre, had fallen throughout the Centre, and standing ponds and lakes of water were commonplace on either side of the road, especially north of Coober Pedy. The old dirt highway to Alice Springs would have been an impassable mire. We continued to make good time from Coober Pedy to Alice Springs, and arrived through Heavitree Gap in the MacDonnell Ranges at Alice Springs late in the afternoon. There we checked into a caravan (trailer) park, where we had booked an onsite van, by far the most economical accommodation for travelers in Australia. Our van was just across the street from the conference center where the Mammal Society meetings were to start the next day. While I attended the meetings for the next couple of days, Barb and Ian toured the town and nearby surroundings. A barbecue was held one evening at Simpson Gap just out of town, in a lovely sandy creek bed with River Red Gums all around. Ian was delighted when, during the proceedings, he and some other children discovered several small, desert frogs (*Limnodynastes spenceri*) hopping about on the damp sand. The following afternoon Barb and Ian returned to Simpson Gap to observe Black-footed Rock-wallabies (*Petrogale lateralis*) that come out to feed and drink at dusk. On another evening, we were invited for a barbecue at the home of Bill Low, a consulting biologist and excellent natural historian of that region. Bill informed us of a population of Bourke's Parrots (*Neophema bourkii*) that regularly came in to water about sundown or slightly after at a dam on a station just west of Kulgera, immediately north of the Northern Territory-South Australia border. He was certain the station owner would give us permission to net, and a phone call provided us with the necessary permission to visit on our return to Adelaide.

We had arranged with Alan Newsome to be escorted around some of his old bush haunts after the meetings, and spent a couple of memorable days with him doing just that. The first took us north of Alice Springs and then east, on a loop that passed The Garden, a cattle station, where we called in on the Turners, Alan having done a fair amount of work there during his graduate school days. We listened to Alan and Mr. Turner reminisce over a few beers, and then went on to the old gold mines and ghost town of Arltunga (see Kimber 1986, "Man from Arltunga") for a look around before returning to Alice Springs.

In 1980 Alan had written a paper (Newsome 1980) on the relationship between Aboriginal mythology and ecological reality, which attributed a lot of their myths to keen powers of ecological observation and interpretation, passed down through generations by word-of-mouth. Much of the paper was based on Alan's own observations and contacts with Aboriginal elders around Alice Springs while he was working there in the early 1960s. Barbara and I were particularly curious about an ancient totemic site that he had tracked down by combining mythology with his ecological observations, and were quite pleased when Alan agreed to take us there on our second outing. Alan said that it had been a sacred site that only could be visited by male totemic elders of the Aranda Tribe. Since there were no surviving members of that moiety, he could show it to us without infringing on tribal secrets. None of us could arrange the loan of a 4-wheel drive, so we set out in the trusty "Silver Bullet" with five of us packed in: me, Barb, Ian, Alan, and Peter Frappell. The site is northeast of Alice Springs, at the northern base of the MacDonnell Ranges; accessible through a cattle station with permission generously granted when we stopped in. I quickly proceeded to get us stuck in fine sand from which we eventually extricated ourselves. However, sand is almost a pleasure, compared to sticky mud and clay. Alan then guided us down a network of increasingly dim dirt tracks. At last, he said, "I think it's over there". "But there's no track", I protested. Alan assured me that the scrub wasn't very thick, and the surface was firm, and that "she'll be right!" After dodging shrubs and Spinifex clumps for about half an hour we came to a windmill and dam, and Alan said "It's just over there". He pointed across a hard surfaced sandy flat dotted with Spinifex clumps, to a rocky hillside, and a small canyon with a conspicuous white mineral apron near its bottom. I drove as close as I could get before the Spinifex got too thick, and then stopped. Shortly before stopping we came across a magnificent male Red Kangaroo (*Macropus rufus*) that stood erect, eyeing us (all six plus feet of it), while Barbara photographed it. Seeing this animal must have been symbolic because the place we were seeking is an Aboriginal Red Kangaroo site called Krantji, meaning neck of the Red Kangaroo (Newsome 1980). We followed Alan up to the site on foot, dodging the sharp spines of the Spinifex Grass (*Tnodia clelandii*) clumps. The white apron turned out to be a flat rock surface coated with limestone that precipitated from water seeping over it. To its side was a very large boulder under which was a small, horizontal cleft; on its floor was a soak of damp sand. Ian found a partial skeleton of a small wallaby in the cleft, and digging in the sand at the bottom of the cleft revealed water to be present. According to the myth Krantji was a totemic site of the Aranda people, where the Red Kangaroo Dreamings were celebrated, for it was there that the kangaroo ancestors stopped for replenishment from the soak during their overland wanderings from one totemic site to another. It was a site of major ceremonial importance, and we felt honored and a little more than humbled to see it. We found it especially interesting to witness the correspondence of the physical features of the site with the Aranda myth that led Alan to it (as translated by Strehlow 1971):

Hail to thee, Krantji, mother of men!
Be fruitful in the ancestral embrace,
Filled with game for the use of men!
The crutch fat is gleaming white,
The crutch fat is white like sand
The Rock-plate quivers as the avengers arrive:
Our Rock-plate of white, fat, —
Our Rock-plate is quivering, our Rock-plate is quivering and stirring.
Our windbreak home of white fat,—
Our windbreak home that gleams like sand.

Alan was convinced that the flat rock slab coated with white limestone was the "Rock-plate of white fat... that gleams... like sand" of the Krantji myth; upon seeing it we agreed.

As it was well past noon, and it had taken better than half a day to get there, we figured that we had better be on our way so that we could reach pavement before darkness. Alan remembered that a track went off from near the dam we had passed, and then headed back to intersect the paved road much closer to us, so that we wouldn't have to back-track through the sand trap and station to reach pavement. Although a bit dubious I followed Alan's directions and, sure enough, we came on a vague track that headed in the right direction, so we followed it. The track got increasingly vague and finally ended abruptly at a barbed-wire fence. Alan pointed down a rocky wash, and said that it used to follow the wash through some rocky ridges to the paved road. Certainly no one had been through there since the last time the wash ran with water, and it appeared impassable, even with the trusty, two-wheeled drive "Silver Bullet". Had we four-wheel drive we might have proceeded, but, as dusk was rapidly approaching, we turned around, retracing our steps as quickly as possible, ducking most of the Spinifex clumps, but running over a few. At the dam we headed cross country, taking the shortest course to the station track, dodged the sand trap, and made it to the paved road back to Alice Springs just as it got dark. Later, after our return to Adelaide, when I was having the "Silver Bullet" serviced, the mechanic remarked "half the Spinifex from the Northern Territory was lodged under the vehicle".

Our last couple of days in Alice were spent as guests of Steve Morton and Faye Alexander, and their children Josh, George and Carrie. Steve and Faye had spent 18 months in the late-1970s in southern California while Steve was doing postdoctoral work with me on desert rodents, and Josh was born during that interlude. We had last seen them at Jabiru, in the "Top End", east of Darwin in 1983, where Steve was a government biologist conducting vertebrate censuses in the wet-dry tropics (six months of wet monsoons, six months of no rain). Since then, Steve had been appointed as an ecologist at CSIRO's Centre for Arid Zone Research in Alice Springs. It was very good to catch up with them, and to meet Carrie, who hadn't been born when we visited them in 1983. At the Morton-Alexander household, we arranged for our trip south, reconfirming our visit to the station near Kulgera to look for Bourke's Parrots. Since we would be returning to Alice before our final departure back to Adelaide, we left items unnecessary for our collecting trip with the Morton-Alexanders to fetch later.

It rained nearly all night before our scheduled departure for Kulgera, and we got off to a late start, optimistically hoping the rainfall would be only local. At first, the rain was gentle and sporadic, but it became harder and more regular as we went further south. At Kulgera, it was raining rather hard, certainly not inviting camping weather nor did the roads look particularly compatible with our two-wheel drive "Silver Bullet". We decided to take a room overnight at the hotel in hopes that things would look better in the morning, but soon discovered that the daughter of the station owners, who had given us permission to net parrots on the property, was getting married that day, and the entire Hotel was booked solid to accommodate relatives and guests. There was no room at the Inn. We decided to have some refreshments at the hotel while we rethought our itinerary. The hotel was filled with wedding guests who were waiting to be taken by 4-wheel drive vehicles to the station. Obviously the roads required 4-wheel drive to get through and the rain was not letting up so we decided we would not be able to get into the station the next day, and we would need to make alternative plans. Before leaving Adelaide I was told that Mabel Creek Station, west of Coober Pedy, might be a good place for Bourke's Parrots and Budgies, so we decided to stop by there on our way back to Adelaide. Since we had to return to Alice Springs to collect the rest of our things, we headed off without delay. To reach Alice would require a long rainy drive in the dark, so we elected to find cheap accommodation along the way. At Erldunda, the turnoff to Uluru (Ayers Rock), the "budget" rooms were overpriced and under equipped, but available, so we decided to go on down the Lasseter Highway toward Uluru to see if anything was available. Nothing was. Swallowing my Scottish pride, inherited from my frugal father who would have camped in the rain anyway, I agreed to go back to the overpriced "budget" motel, and hope that a room was still available. Luckily, one was and, retrospectively, probably worth the price!

We returned to Alice, were welcomed by the Morton-Alexanders, and a call to the owner of Mabel Creek Station, Shane O'Connell, resulted in an invitation. We repacked our things and got off to an early start for Coober Pedy and Mabel Creek Station, hoping to leave the rain behind. I phoned the station owner at Kulgera and explained that we would not be coming, and he understood fully. By the time we returned to Kulgera the rain had stopped, but all of the dirt tracks still would have been impassable for the "Silver Bullet", so our choice had been the right one. We reached Coober Pedy by evening and, feeling budget minded, found a low-priced, rather grungy aboveground motel, but made up for it with a return visit to the nice underground restaurant we had discovered earlier, for dinner. The next morning, after fueling up, we left for Mabel Creek Station, about 40 km west of Coober Pedy, and just off the old dirt highway we had traveled in 1983. The owners at Mabel Creek greeted us warmly, and said we were welcome to try our luck most anywhere on the station. While we were being shown the homestead dam, the owner, Shane, commented in a typically understated manner, "the flies are a bit sticky today!" He was referring to the bushflies of which several thousand were on our backs, with at least another thousand around each of our faces, trying to get into our eyes, ears, nostrils and mouth. I'll tell you more about them later. Shane directed us out along the old highway, near where there were several waterholes. He also advised us to look for his hired hand who was driving a road grader out there, and who knew about birds. The hired hand, who did seem to know about birds, thought that nearby Government Hole would be our best bet, because it usually attracted lots of parrots of different kinds, and we could reach it easily with two-wheel drive; he said there were other good places, but they all would require four-wheel drive, what with the recent rains. The weather by this time was quite spectacular, with white puffy clouds in an otherwise deep blue sky.

A short time later we arrived at Government Hole (see fig. 11), a beautiful large, deep billabong, one of a chain of several along a non-flowing creek channel, with the banks lined with huge, spreading Coolibah Trees (*Eucalyptus microtheca*). As this was a sheep station, Ian thought this was a perfect setting for "Waltzing Matilda". It really was a beautiful place, particularly in contrast to the desolate, starkness of nearby Coober Pedy. We selected a flat sandy area under several Coolibah Trees, and well above the present water level in case of sudden flooding. As we set up our camp we were relieved to see quite a few parrots cavorting about in the trees, and flying overhead: Budgies, Galahs, Little Corellas, Cockatiels, and Mulga Parrots (*Psephotus varius*), and lots of Zebra Finches. We knew though that there was water everywhere due to all the recent rain, and that netting wouldn't be easy. This notion was reinforced by the observation that all of the birds we saw were high in the trees or air, above the reach of our nets.

The first couple of days were warm and beautiful, with billowy white clouds drifting through the dark blue sky. We set nets here and there along the banks of the billabongs, checking them periodically, but nearly all parrot flights were at tree height, not net height, and the birds were not coming in to water. Instead, they used the trees for roosting and preening, flying periodically out toward the grassy flats away from the billabong, probably to feed. We placed one net between some smaller trees away from the water, where we had seen a few Budgies and many Zebra Finches, and proceeded to catch only several of the latter. The Budgies avoided the nets, and elected to sit in trees just above our camp. Two pairs, in particular, demonstrated their complete indifference by copulating repeatedly just over our heads, as if to say "we have better things to do than to fly into nets!" Periodically we would move the nets to what might be a better location, but

Figure 11. Government Hole, a billabong on Mabel Creek Station, near Coober Pedy, South Australia, 1989.

pretty much to no avail. We finally managed, after two full days of netting, to catch one Budgerigar, which we kept, and two Mulga Parrots, which we held temporarily, in the event of catching more. Between checking nets Ian and I invented a great game that we labeled "Beer Bowls". This was our version of "Lawn Bowls" (lawn bowling), in which empty beer cans were placed in a triangular array on a flat, smooth part of the billabong bank near camp. We took turns bowling Paddy Melons at them. These are the round, gourd-like fruits of the Wild cucumber (*Cucumis myriocarpus*), about the size of baseballs.

On the third day, the weather changed, and we could see large dark clouds building up in the northwest, a little late for summer monsoons as it was May. As the day proceeded, the sky to the northwest got darker and darker, and we could see flashes of lightning lacing through it; it was apparent that this weather front was moving toward us. During this period, we somehow managed to net a Little Corella, which we released, and a lone Galah, which we kept. As the storm approached closer, we began to prepare by having an early dinner, and securing the birds, our food and other dry goods in the camper shell. Our cheapo tent did not have a rainfly, but fortunately, we had purchased a large blue plastic tarpaulin with which we quickly fashioned a rainfly over the tent, securely fastening its ends to the ground with rope and extra tent stakes. All this time the storm was getting closer, with lightning, followed ever more closely by booming thunder, flashing through the black-purple clouds. Barbara and I knew that we were in for something unpleasant.

Trying not to show our alarm to Ian, we ducked into the tent just as rain began to patter down. Then the wind began to blow, and the rain intensified, and lightning streaked through the sky all around us, illuminating the otherwise dark world. It was still early evening, yet everything was dark. Over a two-hour period, the storm intensified still further, with the wind coming in periodic, powerful gusts. We feared that the tent stakes would not hold against the wind, and that the tent would soon collapse. Not wanting to spend an even more uncomfortable night in the cab of the "Silver Bullet", I decided to go out into the storm to maneuver the vehicle perpendicular to the wind direction, making a windbreak for the tent. In so doing, I became completely soaked in the driving rain, but the trick worked and the "Bullet" took the brunt of the wind, protecting the tent behind it. I ducked back into the tent, removing as much gooey mud from my bare feet as I could, peeled off my wet clothes, dried as best I could, and dived into our double sleeping bag, damp and cold. Barbara and the down sleeping bag soon warmed me up, and we lay there listening to the continuing fury around us, but much relieved that the tent was no longer straining on its stakes even though the wind continued unabated. Ian, somehow, had managed to go to sleep, secure and snug in his little down bag and his peaceful dreams.

The rain continued to flood down. Fortunately, the tent was on a slightly inclined bank so that the water ran off toward the billabong. When I got up to move the vehicle I also took a shovel, and quickly scooped out ditches upslope from the tent, to divert runoff water away from it. These worked well, and very little water pooled up under the tent, and the rainfly shed water just as we had hoped. We remained quite snug, but for a while I was plagued with the secret fear that all of that rain would cause flooding in the creek, raising the water level up to our tent. We had placed the tent well above the billabong surface, but we certainly didn't anticipate the kind of rain we were experiencing. And no telling what was happening further upstream in the watershed. I was reassured by hearing no water flowing, and we managed to sleep fitfully, awakened periodically by flashing lightning and booming thunder, frequently coinciding. We were also reassured by having placed the tent in the open, rather than under a tree that might have been struck by lightning, or lost a limb under the onslaught of the wind. By midnight, the storm had passed by, and we drifted off to sleep with the rain, lightning and thunder moving further and further away. Ian had slept comfortably through the whole ordeal.

The next morning we awoke to a cold but sunny morning, with a much gentler breeze still blowing out of the northwest. The ground was still damp, but rapidly drying, and we managed to find enough dry firewood under trees to get a campfire going, and all was quickly returned to normal. Our slightly soggy sleeping bags were draped over limbs for drying. The parrots we had captured were fine and healthy. The weather to the northwest still looked somewhat unsettled, so we decided that it would be best to spend the day drying out our belongings, reassembling and packing our gear, and depart early the next morning, rather than risk another storm. Although we had only one Budgie and one Galah to add to our study, our time had pretty much run out, and we would have to figure out another means of obtaining the birds that were necessary for completion of the study.

The storm did provide one temporary benefit, though, that of reducing the number of bushflies that had been plaguing us previously during all the daylight hours. Bushflies (*Musca vetustissima*) are small (i.e., about one-third the size of the common House Fly) flies that occur by the millions throughout the interior of Australia, and they thrive particularly during hot and warm weather. They are active only by day, and are attracted to moisture and anything that moves. Hence, perspiring humans are perfect targets, and a warm day will often find swarms of them around one's face. Fortunately, they don't bite, but their mere numbers and persistence are enough to drive a person to irrational acts, such as swearing violently and flailing one's arms wildly about in the air. Fortunately on a day that we had gone into Coober Pedy to replenish our ice, Barb purchased fly nets for each of us; when draped over the face from the brim of a hat, they at least moderated the problem, reducing the number of flies in the face to a tolerable level. The only real relief from them is after the sun goes down, when they retreat to the bush and become inactive. Even then, after a particularly "sticky" day, you could still imagine that they were crawling around on your face. Barbara lay in the sleeping bag one night, after a dreadful day of constant fly bombardment, with the imagined, but realistic tactile sensation that they were still creeping about on her. Often, when awakening in the morning, the rising sun would be accompanied by a buzzing outside the tent, and from within you could see an accumulating mass of the little black buggers, waiting to make your life miserable as soon as you emerge; you can imagine that eating on a sticky day was a tricky ordeal!

While waiting for birds to fly into our nets at Mabel Creek, we had plenty of time to explore the local area. Our camp was right next to a series of chert outcrops, a flint-like mineral that was used widely by Aboriginals for manufacturing simple rock tools. Flakes of chert covered the ground, some of them with nicely shaped cutting edges. Ian was most impressed, particularly when he cut his finger on one that he had found. An ideal place like this with water, edible animals and plants, trees for shelter and firewood, and chert for tool-making, was likely used by generations of Aboriginals; hence the enormous numbers of flakes and artifacts on the ground. Our exploring and grimy bodies also led us to venture into some shallow pools below the main billabong. These were filled by seepage of water from above through gravel and sand, and the water was nearly clear and warm, ideal for bathing. Ian loved these pools, because they also had an abundance of creatures. Not only did he find polliwogs of unknown frog species, but also he found numerous Shield Shrimp (*Triops australiensis*). These curious, very primitive-looking crustaceans occur especially in inland freshwater habitats that periodically dry up. Unlike the Inland Crab, this species over-droughts as minute, drought tolerant fertilized eggs, the adults dying as their aquatic habitat dries up (Williams 1968). Those that Ian collected were up to two inches in length, resembling fossil trilobites. Ian also found at least two species of Clam Shrimp, of unknown identity, one large (about one inch), and one smaller. These are equally curious shrimp-like animals, with the head and body enclosed within a bivalved fibrous shell. They also are said to over-drought as desiccation-resistant eggs (ibid.), rather than as adults.

The morning of our departure from Mabel Creek Station happened also to be Barbara's birthday, which was celebrated with an early camp breakfast decorated with wildflowers that Ian had picked, completion of packing, and our departure, with the intent of driving south to Port Augusta for a motel room, and a good restaurant for a birthday dinner. On our way out, we stopped by the station homestead to thank the owners, where we found mum and children in the adjacent schoolhouse, engaged in "School of the Air". Then we were off to Coober Pedy for fuel, and thence to Port Augusta, which we reached about dusk. We managed to get the last room for three in a decent motel, into which we snuck our birds for safekeeping. The motel manager advised us that the best restaurant in town was at a Hotel, but that a wedding reception was being held there that evening so it wasn't available to the public. The next best was at the RSL Club (Returned Service League Clubhouse), where we would be welcome. Although it was rather like dining in a gymnasium, we did have a pretty good meal, and a roving photographer came by for a birthday photograph. All this was quite a contrast to our camp at Government Hole!

The next morning was an easy trip back to Adelaide where we dropped the birds by the University, before going to our home to get settled again. It was a long and memorable trip, but one that resulted in only two birds, one each of the two most common parrots in Australia: Budgerigars and Galahs. With only seven weeks left before our departure from Adelaide, we still had a long way to go with our measurements.

In our absence, Russ had the Red-rumps adjusted to their minimum water ration, and so we whipped them out in short order. A few more phone calls resulted in the loan of four more Budgerigars from Cleland Wildlife Park, operated by National Parks and Wildlife, near Adelaide. These, together with our three wild-caught Budgies provided an adequate sample. A few more phone calls led us to a licensed bird-breeder near Adelaide, who was willing to sell us eight captive-bred Bourke's Parrots at a very reasonable price. Although we were uncertain of how natural physiological traits would hold up under several to many generations of captive breeding, this species was critical to our study, and we took them gratefully. This would then provide us with adequate samples of five species of parrots that spanned a broad size range, so we decided to confine our measurements to these during the MacMillens' remaining weeks in Adelaide, while deferring the Galahs to a later time if capture and Russ' schedule proved possible, which they did. We just managed to finish all of our measurements about a day before our scheduled departure by working both day and night, and without further meanderings except those of the mind.

As is typical of most experimental work, our approach and methodology benefited from a lot of prior work by others. Especially important were the observations of my mentor, George Bartholomew, in his earlier studies of small seed-eating birds. He (Bartholomew 1972) was the first to affirm that the two major components of water balance in these birds, metabolic water production (input) and evaporative water loss (output), become uncoupled at lower environmental temperatures, with the former increasing and the latter decreasing as temperature decreases. Bartholomew (ibid.) stated further that for each bird some low temperature exists where the two components are equal, and water balance is achieved without drinking water, with water surpluses resulting at still lower temperatures. Based upon these observations, those of others, and my own on water-independent seed-eating rodents, I was able to formulate a model of water economy in granivorous birds that would predict the circumstances under which a seed-eating bird could become independent of drinking water (MacMillen 1990).

The observations and assumptions upon which the model is based are: 1) some small seed-eating birds thrive in the laboratory under moderate conditions, without drinking water; 2) they apparently derive all the water they need from the small amount absorbed in the dry seeds they eat, together with a larger amount produced as a byproduct of metabolism (metabolic water); 3) larger seed-eating birds require drinking water in the laboratory; 4) small birds are independent of drinking water at moderate and lower temperatures, but not at higher temperatures; 5) in the water budgets of these latter birds, the small amount of water that is absorbed in seeds that are eaten is about the same as that lost in the excreta; the largest avenues of water gain and loss are metabolic water and evaporation from lungs and skin, respectively; 6) the rate of evaporative water loss for a given bird at moderate to low temperatures remains stable or decreases within this temperature range, while the rate of metabolic water production increases as temperature decreases within this same range; and, 7) since the water absorbed in consumed seeds is equivalent to that lost as excreta, simultaneous measures of metabolic water production and evaporative water loss should provide an accurate estimate of momentary state of water balance.

From these model components we predicted that: 1) within a size spectrum of granivorous parrot species, smaller parrots should be much more economical in their water requirements than larger parrots; 2) for each parrot, regardless of size, there should be some temperature at which metabolic water production equals evaporative water loss, resulting in water balance; 3) this temperature at which metabolic water gain equals evaporative water loss (the temperature at which water balance is achieved) will be higher in small parrots than in larger ones; and, 4) therefore some of the smallest species of granivorous parrots should approach drinking water independence under some of the low temperature conditions they encounter in their habitats.

Our experimental protocol enabled simultaneous measures of evaporative water loss and oxygen consumption at a variety of environmental temperatures (see fig. 12); our oxygen consumption measures allowed accurate estimates of metabolic water production while the birds were assimilating carbohydrate-rich seeds. With this protocol, we could then determine the relationship between metabolic water production (MWP) and evaporative water loss (EWL) for individual parrots across the complete size range of our experimental species (ca. 27 g to 391 g), and at various temperatures. These data provided interpolations of the temperatures at which MWP = EWL, and the achievement of water balance. Our parrots conformed very closely to the expectations of the model, as we found that smaller parrots achieved water balance without drinking water at higher temperatures than larger parrots. In fact, the two smallest species (Budgerigar and Bourke's Parrot, 27 and 39 g) could maintain weight indefinitely without drinking water at moderate and lower laboratory temperatures and achieved water balance (MWP=EWL) at temperatures encountered at least in winter in the field. The four larger species were progressively more dependent on drinking water. We also found that, within a given parrot species, individuals were more efficient at water regulation at night than during daytime, and that a period of tolerable water restriction resulted in even greater efficiency both day and night. Our studies are detailed in MacMillen and Baudinette (1993).

Figure 12. A Bourke's Parrot inside a respirometer chamber at Flinders University of South Australia, 1989. The chamber is equipped with ports for the introduction and elimination of air, and for a temperature probe.

Piecing these laboratory observations together with field observations and what we know of parrot behavior, suggests the following scenario: 1) parrots in general are strong fliers, with larger ones having greater flight ranges than smaller ones; 2) at least in the Channel Country, where rainfall is predominantly summer and monsoonal, drinking water is likely continually available at waterholes within the flight ranges of all species during seasonal periods of higher temperatures, when drinking demands are greatest; 3) During winter periods of drought and lower temperatures, surface water is more dispersed, still within the flight ranges of larger parrots, but not necessarily of smaller parrots; 4) during these same winter periods food in the form of seeds may be equally widely dispersed as water, but in very different places, such that energy may be acquired in one place and water in another; 5) the more mobile larger parrots can readily fly from feeding sites to watering sites, easily meeting all of their nutritional needs; 6) reduced water availability during drought in the less mobile smaller species will result in increased water regulatory efficiency, providing these birds with more time for meeting energy needs while feeding; 7) night-time adjustments increase still further the water regulatory efficiencies of these small parrots that typically are inactive at night; and, 8) at least during the cooler winter months, while operating at maximal water regulatory efficiencies, small parrots such as Budgerigars encounter environmental temperatures that result in positive water balance (MWP=EWL) for most of the 24 hours; the slight water deficit incurred during the few daytime hours of higher temperature may be readily offset by the longer night-time period of metabolic water surplus, resulting from the lower nocturnal temperatures. Thus, it is theoretically possible for small parrots such as Budgerigars and Bourke 's Parrots to remain for at least several days exploiting a rich seed supply during the cooler winter months, without having to disrupt energy acquisition and employ energetically expensive flight to seek water at a distant site.

Following the completion of most of the experimental work, we departed our temporary home in Adelaide and Flinders University. Our return to the States took us via Canberra for a short, but pleasant visit with the Newsomes, and a seminar at CSIRO's Division of Wildlife and Ecology to present our preliminary parrot results to an appreciative audience; then to Sydney for a visit to Dangar Island so Ian could see where he was born; and finally on to Fiji for a delightful cruise of the Yasawa Islands, blue skies, warm clear water, sparkling white coral beaches, and some of the best snorkeling in the world over spectacular coral reefs. At nearly six years of age Ian proved to be an adept snorkeler, although he became a bit apprehensive when Barb and I guided him over a hole in the reef on the bottom of which rested a four-foot sand shark. It was then time to face the real world again, well almost! We had carefully arranged our time so that we could spend the month of August at our wilderness retreat on the East Fork of the Salmon River in central Idaho, gearing up gradually for re-entry back into the fast lane of southern California. Unfortunately it didn't work, and our gears were mighty sluggish by the time we returned, but that's another story. Our sabbatical was most successful from every perspective that was important to us, with meanderings that will last in our memories forever.

SUGGESTED FURTHER READING

MacMillen, R.E. 1990. Water Economy of Granivorous Birds: a Predictive Model. Condor 92: 379-392.

MacMillen, R.E. and R.V. Baudinette. 1993. Water Economy of Granivorous Birds: Australian Parrots. Functional Ecology 7: 704-712.

Chapter 6

ALICE SPRINGS AND THE ULTIMATE VISIT TO SANDRINGHAM, 1992

On the banks of the Georgina
And down the Diamantina
To where the grass is greener
Down by New South Wales.
Johnny Stewart's roving
With mobs of cattle droving
His life story moving
Down miles of dusty trails

Contemporary Bush Ballad by Chris Buch

Having stressed the importance in my professional life of serendipitous happenings, I must now confess that the initial drafting of this book was also a fortuitous event. I had planned to spend my 1992 sabbatical leave in Alice Springs with Steve Morton at the CSIRO Centre for Arid Zone Research, conducting a study of the field metabolic rates of coexisting small marsupials and rodents at Uluru National Park, paying particular attention to the roles that rainfall and bushfires play in their population energetics. This would be an ambitious field project for which I had requested funding from the U.S. National Science Foundation. I had had the notion of this book in the back of my mind for some time, but without any concrete plans in the near future of undertaking its writing, until N.S.F. rejected my proposal for funding. This opened the door to a much more relaxed sabbatical leave, allowing me to explore my creativity in a different way, and I must therefore acknowledge N.S.F. for denial of support in making this book possible, and also acknowledge the serendipitous manner by which it proceeded.

My 1992 leave from UCI was for the spring academic quarter, allowing a four-month visit to Australia with our summer month of July tacked on. Barbara, Ian, and I packed up our things and departed for down under in late March 1992, stopping off en route in the Canterbury district of South Island, New Zealand for some sightseeing. We arrived in Melbourne, Australia in early April, where we were met at the airport by my good mate, Tony Lee, and his companion Janet Doyle. Tony had retired from the faculty of Biology at Monash University and was embarking on a nature tour business. Our two or three days in Melbourne were spent catching up with Tony, and visiting automobile agencies in search of a used vehicle to purchase. Against Barb's and Ian's advice I purchased a 1987 dual cab Nissan pickup, rather than the four-wheel drive we should have bought. This vehicle was too conventional to dignify immediately with a name, which would come later and which I even now regret. We then repacked our suitcases and headed for Adelaide where, with Russ Baudinette's help, we would intercept the air freight shipment of our Macintosh computer and camping gear. This time we had not included Mexican salsa, as we thought we might (barely) survive the four months without it.

Tony had arranged for us to stop over at Whimpey's Little Desert Lodge on the Victoria side of the Victoria-South Australia border, to rendezvous with Terry O'Brien, also an Emeritus Professor of Biology from Monash and Tony's partner in the nature tour business. Terry was there conducting a tour of high school students from the International School at Kuala Lumpur, Malaysia, who were on spring break. Whimpey's was a delightful place and Whimpey a delightful character, who, by coincidence, was celebrating his birthday the evening we arrived, replete with multiple bottles of good Aussie champagne. The Little Desert Lodge was a renovated Technological College purchased by Whimpey. The surrounding area that he owns is old farmland that he is gradually restoring back to native Little Desert shrub lands, abutting against Little Desert National Park. Whimpey is an excellent self-trained naturalist and photographer, and is developing a nature reserve well worth seeing, together with very comfortable dormitory accommodation and excellent food. Ian was quite taken with the tame Grey Kangaroo joeys (either Eastern or Western Greys, *Macropus giganteus* or *M. fuliginosus*, respectively; I was uncertain which, as they both occur in the area) and Emus (*Domains novaehollandiae*) that wandered about the place.

Our stay at the Little Desert Lodge was a brief one as we were to meet Russ, Kaye and Emma Baudinette the next evening in Adelaide, a day's drive to the west. While in the process of departing Whimpey's I discovered that our vehicle had a dead battery, the first of several indignities it managed to perpetrate on us. A jump-start from an agreeable chap delivering hay got us going, but the poor vehicle was to suffer thereafter from its name of "Wally". In Australian "Wally" is a person who seldom does anything right! This was Barbara's and Ian's idea, and I am sure that it was meant as much for me for purchasing it as it was for the vehicle; but "Wally" it became. Russ had our shipment at his house on our arrival and all was in order. We had planned a nice weekend reunion with the Baudinettes, which was spent on a houseboat on the Murray River near Mannum, just east of Adelaide. Two very peaceful days on the river allowed us to catch up with our busy lives, to relax with the children, and to wonder at the myriad birdlife on the river. We were particularly taken with the scores of Sulphur-crested Cockatoos (*Cacatua galerita*) that inhabited holes in the bright red cliff face along the river. Their stark white color and bright yellow crests contrasted vividly with the background of red rock. Ordinarily these birds nest and take shelter in holes in gum trees, but these seemed quite content in their cliff-dwellings.

After a restful visit in Adelaide, we were anxious to start the two-day haul up the Stuart Highway to Alice Springs, so that we could arrive a day or two before the Easter Holiday, when the whole country comes to a grinding halt. Besides, we had been living out of suitcases too long and were anxious for a little more permanency. Ian was eagerly looking forward to school in Alice Springs and to escaping continuous adult company and conversation.

The drive up the now-sealed Stuart Highway was uneventful, with a stopover for the night at an over-priced motel at Glendambo, in central South Australia. Dark storm clouds loomed all around us threateningly, following us first from Melbourne and then from Adelaide. I have never failed to bring rain to the dry Centre. We had only a plastic tarpaulin covering our luggage and computer in the bed of the truck, and I was uncertain whether this would be adequate protection if we were really caught in a downpour. It was a long haul from Glendambo to Alice Springs, particularly since we stopped for a look-around and some opal shopping in Coober Pedy, so it was nightfall before we arrived in Alice. Somehow we avoided serious downpours of rain, although thunder showers were occurring all around, and we arrived in Alice Springs unscathed, about 7:30 PM, just in time for a rainy barbecue dinner at Steve Morton and Faye Alexander's home. After a little chatting with the Morton-Alexander tribe, Steve showed us to our home in Alice, owned and rented to us by the CSIRO unit we would be working with. It was a modest, but comfortably furnished house, and Faye had graciously stocked the fridge for us to get us through the first day or two. It was good to be home.

Thursday 16 April preceded Good Friday, and would be the last day of school until Ian started the following Tuesday. As we learned that we would have our option of two nearby schools, we felt it was important to visit each, and to let Ian choose his school. On Thursday morning we visited both Braitling and Bradshaw Primary Schools, in Alice Springs' Public Schools System, and Ian chose Bradshaw where he was assigned to Lucy Skoss' third grade class, having just completed 2/3 of second grade back in California. It was the right choice, the transition was quick and easy, and Ian flourished there under Mrs. Skoss' tutelage. Additional trips on Thursday to the bank, Telecom (the telephone system), and the market got us nearly fully functional, and we prepared for Easter and my birthday, both coinciding on Sunday. This was big number 60; a threat to my youthful psyche that I hoped would be overwhelmed by the other events of the day.

With Ian happily ensconced in school, Barbara and I settled down to a more normal routine on weekdays of writing about past meanderings, and engaging in more meanderings on weekends, such as visits to the nearby bush, the country and western music festival, and the annual Camel races. Although tourism was suffering from a worldwide economic slump, it was still flourishing in Alice Springs and surroundings, with tour buses shuttling people of many nationalities from one place to another, much like column-foraging ants. Alice is very hard to describe, as it is both cosmopolitan and countrified, taking pride in its outback heritage and, at the same time, sporting an alarming display of shops, hotels and motels, and restaurants. I once described it to an Aussie friend who had traveled in the U.S. as a cross between Tonopah, Nevada and Palm Springs, California, much to that person's dismay. Barbara likens Alice to our then hometown of Laguna Beach that grew up as an artist colony and therefore was a magnet to eccentrics. Another description of Alice Springs is that it consists of three concentric circles: the innermost a circle of tourists and tourist attractions, the next a circle of white residents who live and work there, and an outer circle that has no obvious peripheral boundaries, and is made up of the Aboriginal community. To us outsiders who had temporarily invaded one of the circles, there appeared to be little interaction between the three, although none of them is strictly self-sufficient. It is an unusual, but comfortable little community, an ideal spot to experience, reminisce and write about outback Australia and its bush.

The CSIRO Centre for Arid Zone Research where I was writing was ideal for my purposes, as it consisted of an excellent group of natural scientists with specialties in soils, rangelands, vegetation, vertebrate animals, ecology and conservation, all with substantial experience in the Australian arid zone. It was their mission to assess the interrelationships between applied uses of the land (pastoralism, mining, tourism, etc.) and the fragile native biota, with emphasis on sustainable management and conservation.

Their combined knowledge and the libraries of individual scientists and the Centre were readily available to me, and invaluable.

I had commented about the rain following us up to Alice Springs from Adelaide. Well, it didn't pass through, but stayed nearly as long as we did. It rained off and on for the rest of April and for the first three weeks of May, and this was supposed to be the start of the fall dry season in an otherwise desert environment. On Saturday night, 9 May, there was an incredible downpour accompanied by lightning and thunder, that resulted in much flooding and damage around Alice Springs, filling the usually dry Todd River, and blocking the railway and motor vehicle entry to the town through Heavitree Gap, its south entrance. We had just left the Camel races south of town, coming through the Gap to our section of town before the downpour, when flooding commenced. This was the latest in a series of similar personal experiences dating back 26 years, and described repeatedly in this book, that convinces me that either it always rains in central Australia (contrary to rainfall records), or it always rains when I am there.

One of my goals of this sabbatical was to take Ian and Barb to Sandringham Station, to show them the Station and the surrounding Channel Country that had had such a profound influence on my development as a naturalist and meanderer. We had hoped to make our planned return to Sandringham at the end of May, coinciding with the new moon on 1 June, to ensure maximal small mammal nocturnal activity. The continuing rain began more and more to make that goal unlikely, particularly when I was advised by telephone to local police in the areas we would pass through that the roads were impassable even to four-wheel drive. We had arranged for the loan of a four-wheel drive Toyota truck, but only a boat could make the trip as conditions stood by mid-May. Finally, the weather began to break to the north and east of us, toward our destination, but rain continued in Alice Springs and to the south.

We had promised Ian a trip to Uluru (Ayers Rock) to see the monolith, and I wanted to observe the area where I had proposed to conduct the small mammal energetics study. Pip Masters, a former technician at CSIRO who had recently completed a Masters Degree on the small mammal population at Uluru, was there doing some follow-up work with her Masters supervisor, Lynn Day, formerly my Ph.D. student at U.C., Irvine (ah, the web of academic nepotism gets more entangled!). We decided, weather not withstanding, that the period 22-24 May would be the ideal time to combine a trip to Uluru with a shakedown camping trip in preparation for our lengthier trip to Sandringham Station, tentatively scheduled for the following week. We drove south toward Uluru to the tourist center of Yulara, and then followed Pip's "mud map" (a crude map drawn with pencil on paper or traced by finger in the dirt) to her study site. She and Lynn then directed us to their camp in a Mulga (*Acacia aneura*) grove several kilometers down the dirt track. Although there were some clouds overhead, the sky did not look threatening and I was relieved. We found the camp with no trouble, erected our new, yellow three-person tent, into which our foam sleeping pad and three sleeping bags just fit. The tent was placed on slightly sloping ground, above the level of most of the surrounding terrain, in pessimistic symbolism to rain. We had volunteered dinner, which Barb and I began to prepare, as Ian explored the vicinity of camp. It wasn't long before he and Barbara, who had gone off after firewood, came up to me to display something Ian had found. He asked "Dad, is this a woomera?" (a woomera is a wooden spear thrower used by Aboriginals). It was indeed a woomera, lying at the base of a Mulga tree, partially cracked and decomposing. Remembering the opal Ian had found at Coober Pedy, I said to myself "It's either a woomera or a pony, I don't know which!" Ian was aware that Uluru is both a National Park and Aboriginal land, and begrudgingly agreed to place the woomera back where it was found, after photographing it. He understood the reasons, but understandably was disappointed. There also was an abundance of chipped and flaked stones strewn over the ground, indicative of past Aboriginal presence.

Pip and Lynn drove into camp about dusk, after setting 300 live-capture small mammal traps, and readying several sets of pit-fall traps with drift fences. These together, could catch most of the array of small mammals and lizards that inhabited the Spinifex *(Triodia basedowii)* grasslands at Uluru that Pip was studying. After a good camp dinner of grilled chicken-on-skewers, steamed rice, and broccoli, which we all ate with gusto, we sat around the campfire, watching the stars through the wispy Mulga trees. Ian and I soon retreated to the tent for his evening story and tuck-in, after which I went back to the campfire. We noticed that there were decidedly fewer stars, and that heavy-looking clouds were rapidly assembling. After a nibble or two of chocolate and another beer, I excused myself, saying goodnight to Barb, Pip and Lynn, who were clustered about the campfire. On the way to the tent, I put our food box in the cab of the truck, and our other items in the truck bed, under a large tarp, in the event of rain. Almost as an afterthought I took a shovel and dug trenches radiating outward from three sides of the tent, into which water would be shed from the rainfly and flow away from the tent. Ian was asleep by the time I crawled into the sack, and Barbara soon came to bed, too, saying that the sky was getting blacker. It was not long after that, it began to rain, at first gently, and then increasingly harder. We were comforted, by knowing that we had a decent (although unproven) tent on higher ground, surrounded by trenching. It rained all night long, Ian sleeping peacefully through it as usual, and we remained remarkable dry.

Pip and Lynn had spent a restless night with Pip sleeping in the back of her truck under a leaky canopy, and Lynn in her tent, which she had unfortunately erected in a low swale that collected water. They got up early, cold and wet, and went out to check traps. Pip had advised us to pack up our gear and see if we could drive "Wally" out on the now soggy track, as it likely would only get worse. We heeded that advice, throwing our dry gear in the back seat of the truck, and the wet gear in the truck bed, under a soggy tarp, and donning shorts and T-shirts, rather than trying to stay dry in the drenching rain. Besides, if we could drive out, we wanted to see what the traps caught. The clay soil around the camp had become saturated with water down to a depth of a couple of inches. The rear tires of "Wally" had sunk down that far, and only spun as I tried to drive out - I couldn't even get it started to roll. Then, with Barbara driving, and me shoveling ahead of the rear tires down to dry ground, we finally got it rolling, and Barbara drove through the slippery mire to the track, stopping on higher ground, while Ian cheered. Water had accumulated in ruts in the track ahead and previous experience told me that if we could maintain enough forward momentum we could probably get through even with just two-wheel drive. Therefore, I volunteered to undertake the experiment, hoping that Pip could jerk us out with her four-wheel drive truck if we became mired. We were lurching along quite well through sheets of water covering the track, not daring to slow down for fear of becoming hopelessly bogged, when we saw Pip's truck four-wheeling up the track toward us. We frantically motioned her aside, and she just managed to back off the track in time to allow us to pass, stopping only when we had reached a high point in the track, above water level. Barbara remarked that she now understood the descriptions I had written about earlier in this book that earned me the name "Bogger Dick", a name that was just as well left behind.

The main track back to Yulara was much sandier and well graded, providing firmer passage, and we followed Pip along it to help check traps at the remaining sites. She caught two species of dasyurid marsupials, one a tiny, wet, insectivorous Ningaui (*Ningaui ridei*) in a pitfall trap (a sunken plastic bucket into which animals fell), and three larger, carnivorous Mulgara (*Dasycercus cristicauda*) in the live-traps. The Ningaui was new to me, as it had only been described as a species in 1975. I had seen Mulgara before, although I had never seen one actually being trapped; their numbers have dwindled markedly, and they are now considered an endangered species (Kennedy, 1990). Pip is especially interested in the Mulgara, and hoped to study them for her Ph.D. at Sydney University, with emphasis on conservation and management.

It continued to rain as we checked the traps and it was wet, cold and miserable. Barb hadn't brought a rain parka to the sunny Australian desert and donned a plastic trash bag to help keep her dry. Ian thoroughly enjoyed hurdling the Spinifex clumps, until he landed in one, impaling his leg on some of the spiny stems; his hurdling was much better after that. After checking all of the remaining traps, we retreated to the heaters of our vehicles and held a hasty conference, electing to make discretion the better part of otherwise imprudent decisions. Our next move was to Yulara, where we rented an onsite van at the local caravan park for an outrageous price, so that we could sleep out of the rain, dry out and clean up a bit, and cook proper meals, with coffee and breakfast coming first.

Barb, Ian and I then drove down to Uluru (Ayer's Rock) to see what it looked like in the rain, as Barb and I had last seen it under warm dry conditions in 1983. The Rock was even more spectacular this time, as the top was shrouded in clouds, with the reddish-purple sandstone glowing as if from some inner light. Rivulets of water streamed down the usually dry watercourses eroded into the softer sediments. As we drove around it, waterfalls were abundantly present, dropping into what must have been huge potholes worn into the rock below. It was an eerie sight, and it is no wonder that Aboriginals hold it in awe and have dedicated many sacred sites around its perimeter.

The next morning, after a comfortable sleep and warm breakfast in the caravan, we said our good-byes to Pip and Lynn, who were off to check traps while we revisited the Rock to determine whether we might climb it. The weather had cleared measurably, and it had stopped raining before nightfall. Still, we agreed that we had made the right decision in abandoning camp, as we likely could not have gotten our vehicle out for another day or two. The sky was intermittently overcast, but with patches of blue showing through as we headed for the Rock, several kms away. It was open to climbing, and we left the decision to Ian whether we would undertake it, as Barbara and I had used his unborn status as an excuse not to climb 8 1/2 years previously. Of course, he decided, with youthful zeal and exuberance, that we should climb it, and off we went. The route to the top and back is only about 3 km round trip, but with a vertical rise of 348 m (1143 ft). The initial two-fifths of the way is a nearly vertical ascent (or descent), with the remainder intermittently sloping and flat. The steepest parts are provided with chains mounted on pipes anchored into the rock, for hand pulling and emotional security. Ian was off like a flash, while we oldies puffed along behind, admonishing him to stay on the narrow path and in sight. It was a cool and nearly windless day, ideal for the climb, and standing pools of water filled the many depressions at all elevations on the rock. Some of these held the same (or a closely related species) ephemeral Shield Shrimp that we had seen at Mabel Creek Station, near Coober Pedy in 1989. Ian, of course, spied them first, and waited impatiently for us to catch up, so that he could show them to us. Their spore-like eggs must have been carried many hundreds of feet up to these depressions by winds from below or as hitchhikers on birds. The dispersal power of these ephemeral organisms is remarkable. We made it to the top after about 1 1/2 hours and much puffing and sweating, and marveled at the view, along with about 15-20 other individuals. There were spectacular cloud formations, and the intermittent sun shone on Kata Tjuta (The Olgas), causing them to sparkle like jewels set in the distant desert of red sand hills, Spinifex and Mulga. The climb down was fast and easy for Ian and Barb, but I had to favor my weak ankles, cushioning each step with my thigh muscles that became fatigued periodically. I worked my way down slowly, pausing frequently to rest my legs, and came to the final, near vertical pitch. Ian and Barbara had already made it down, and I could see Barb below holding up a cold can of Swan Gold (a Western Australia beer) to lure me down - even worse torture than my aching legs! I finally made it, pouring Swan down my dry gullet, knowing that the amber fluid would eventually filter down to my legs and dissolve the lactic acid that had accumulated there, a little-known remedy shared only by those of us with profound biochemical knowledge! Ian and Barb were nearly unscathed, although all of our leg muscles reminded us of our climb for the next several days.

We then drove over to Kata Tjuta for a look-see and lunch. Barb and Ian hiked over to the base of Mt. Olga, while I elected to stay back with "Wally", and inspect the nearby desert and the backside of my eyelids. I was surprised to see huge numbers of rabbit droppings on the desert floor, indicating invasion by the introduced European rabbit, animals attributed to having caused much habitat destruction, and probable elimination of native species through competitive exclusion. When Barb and Ian returned, Barb and I marveled at the changes that had occurred in the region since our last visit in 1983. At that time, Uluru and Kata Tjuta were reached by dirt track; now all of the roads were sealed. Yulara was just being developed and, apart from one rather primitive tourist resort, the campground at Ayers Rock was the only other option. Now Yulara sports at least three rather posh tourist resorts, and a caravan park/campground, and Kata Tjuta has a paved access road and parking lot. While we were there this time, at least five tour buses with probably more than 200 people came and went during a couple of hours. Barb commented after she and Ian returned from their hike that it was certainly different exploring Kata Tjuta with 200 other people; in 1983, Chris, Steve, Barb and I were the only people there. At least the main tourist center, Yulara, has been developed a tasteful distance away from the major natural attractions, Uluru and Kata Tjuta.

Another change that has taken place in the region since our last visit in 1983 was the "Handback" of 1985, when freehold title to Uluru National Park was returned to its traditional Anangu owners (the Aboriginal inhabitants of that region), who then leased the Park back to the Australian National Parks and Wildlife Service. This has been accomplished with the understanding that the Park will be under joint management by personnel from National Parks and Wildlife and the Anangu, through the Uluru-Kata Tjuta Board of Management, in our minds a truly enlightened approach to natural lands management. The primary aim of their management plan is to ensure that the outstanding natural scenery and other cultural and natural values of the Park are maintained, and that the national and international significance of the Park is not compromised (Creagh 1992). This would be similar to the U.S. National Park Service turning Yosemite National Park over to its Native American traditional owners, and then entering into a joint management agreement with them, something that is about as likely to happen as Dave Foreman (Founder of Earth First!) becoming U.S. Secretary of the Interior.

On returning to Alice Springs, we learned that no rain had fallen there, and that the weather had remained fair and warm. So we set our departure for Sandringham Station and southwestern Queensland for the following Friday, 19 May, and only two days later than our original "ideal" schedule. This would allow Ian most of another week of school, a chance for us to catch up with the laundry and shopping, and to assemble additional gear for a much lengthier trip. Steve Morton and Faye Alexander were particularly helpful in arranging for loan of a four-wheel drive Toyota flatbed truck with canvas canopy, two-way radio, dual fuel tanks (a range of ca. 900 km), and a large tank for drinking water, plus odds and ends of other vital camping gear: tables, ax and shovel, propane stove and lantern, spotlight, traps, etc. As I wanted to sample small mammal populations, particularly of the Fawn Hopping-mouse, I would need proper collecting equipment, which, in this instance, was a large throw net. An Alice Springs welding shop provided me with a one-meter diameter hoop of steel construction rod, onto which Barbara sewed mosquito netting. Kmart (yes, it's there, too!) supplied us with a dozen wide-mouth, plastic specimen jars (labeled sugar, flour, salt, pepper, baking soda, etc.). Food supplies for two weeks were purchased, and calls to the appropriate regional police offices in the Northern Territory and Queensland assured us that the roads were fine and dry. Fortunately, Ian's teacher loved science and valued experiences in the bush, and gave her blessings to his coming absence. All signs promised a great adventure, and we were ready to meander back to the bush.

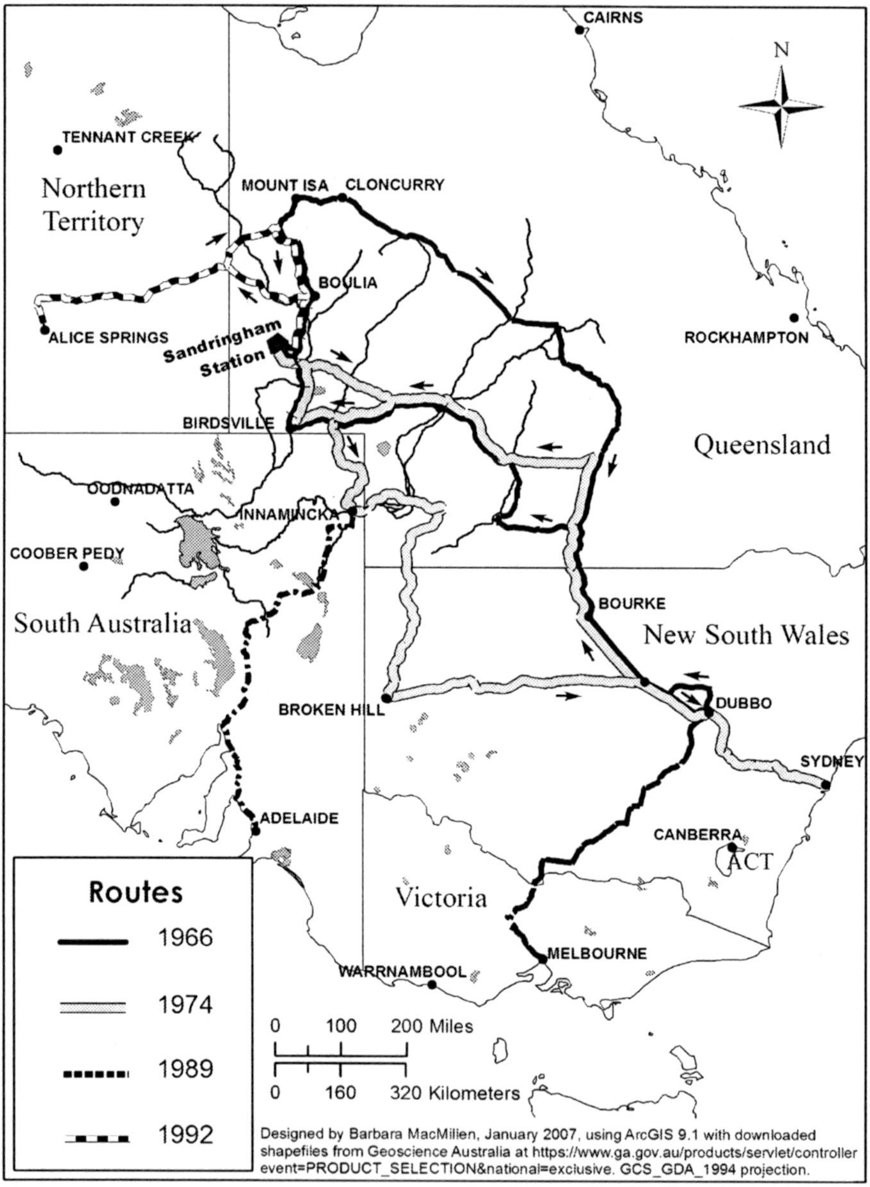

Map 6. Route of our 1992 collecting expedition from Alice Springs to Sandringham Station and return. Our earlier collecting routes are also shown.

We were off to a usual (for the MacMillens) early start by about 10:30 AM, stopping first at the north end of town to fuel up at the "truck stop", where diesel was six cents a liter cheaper than in town. With the amount of fuel our two tanks held, that resulted in a savings of about six Aussie bucks! Barbara and I had traveled in earlier meanderings on the roads between Alice Springs, Tennant Creek, and Mt. Isa, which were the sealed routes into the Channel Country where Sandringham is located, and we wanted to take a shorter, more direct route through country we hadn't seen before (see map 6). The most accessible of these is the Plenty Highway, which is largely an unsealed track heading east to the Queensland border, starting about 60 km north of Alice Springs. This then branches into a northeast track via Urandangi, Qld. to the Diamantina Developmental Road leading south toward Sandringham; a shorter option is along the Donahue Highway extending southeast to intersect the Diamantina Developmental Road near the country town of Boulia, Qld., still north of Sandringham. All of these roads were now dry and safe, although we were advised by the Boulia police to take the longer track up through Urandangi, rather than the Donahue. The weather that first day was clear and mild, and the road was excellent, still damp from all of the rains and much of it recently graded, so little dust. We stopped for lunch just having run out of sealed road near Harts Range Aboriginal community, along the Plenty Highway. At Jervois Station (a working cattle station), the last fuel stop we would encounter along the Plenty Highway, we stopped for an ice cream and to top off the fuel tanks. Between Harts Range and Jervois, a sleek red Dingo crossed the track ahead, stopping to look at us momentarily, before loping off. Just beyond Jervois, we saw two Red Kangaroos, a female and a joey, scampering off through the scrub. Toward evening, we came to a fine, dry creek crossing (Arthur River), and decided to stop there to camp. We drove along an obscure track about 1 km off the Plenty to an ideal campsite on a bluff overlooking the dry, sandy riverbed, with huge, spreading River Red Gums looming overhead (see fig. 13). Our gear was such that we had to unload practically everything before setting up camp, as nearly all of it would be used. While Ian explored the riverbed, we set up the tent, sleeping gear, tables, stove, lantern, and got a fire going to barbecue, all within less than half an hour. As dusk arrived, so did hundreds of Little Corellas, the lovely white cockatoos that Russ Baudinette and I had included in our parrot study. They flew nervously about, squawking loudly at our presence, alighting first atop one gum tree, and then flying up suddenly as if startled, to alight in another. Another commotion started when 6-8 Red-tailed Black Cockatoos (*Calyptorhynchus magnificus*) attempted to land on the gum tree just opposite our camp; they were immediately mobbed by several dozen Little Corellas, and chased off. Red-tailed Blacks are magnificent big birds about the size of Red-tailed Hawks, with glossy black heads and bodies, and with bright reddish-orange panels in their otherwise black tails. As the birds settled down to roost, we sat down to an excellent dinner (again!) of grilled, skewered chicken breast, steamed rice and broccoli, all of us feeling quite pleased with ourselves, that we had outlasted the rains. It was a beautiful, cool night with the milky way clouding the sky overhead, and the Southern Cross looking like the image on the Australian flag: a perfect night in the Australian bush.

Figure 13. A perfect campsite, on the bank of a dry riverbed with River Red Gums looming overhead.

The next morning we awakened to the raucous cacophony of Little Corella cries, as they once again proclaimed ownership to that part of the country. It was a lovely cool, clear morning with no flies, what a blessing. We commented that bushflies are not things that you miss when they are absent, but you sure know it when they are present. As we prepared and ate breakfast, we saw many pink and gray Galahs, and a few Budgerigars fleeting by like small green flashes. We knew we were "out bush" when we heard the plaintive cry of the Australian Crow (*Corvus orru*), that starts off with an almost conventional caw, and then trails off with a series of slower and lower notes, ending with a low, drawn-out gurgling sound.

We packed in a rather leisurely manner, stopping periodically to take photographs. After stuffing his sleeping bag, Ian was given leave to engage in his favorite preoccupation, exploring. This activity has the usual and joyful complication of his wanting to share something interesting that he has found, just as you are in the middle of some packing maneuver, along with adding a few rocks or bones to his belongings. We really had no fixed schedule, and our time was ours. We managed to leave this ideal camping spot about 11:00 AM, and continued west along the Plenty Highway, named we supposed because there was plenty of nature, exactly what we were after. It was certainly plenty far away from Orange County, California and the fast lane!

A short distance down the road a mob of Wedge-tailed Eagles (*Aquila audax*) flew up from the side of the road, where they had been feeding on a recent road-killed Red Kangaroo doe and her yearling joey, likely killed by a passing vehicle the previous night. Invariably the Wedge-tailed Eagles that are seen are associated with road kills, as they seemingly feed preferentially on carrion. One wonders how the introduction of the motor vehicle and domestic livestock, which often end up as dead carcasses along the roads, has affected the populations of Wedge-tailed Eagles in central Australia. We certainly saw them in very large numbers. We soon crossed the border into Queensland, first heading north and then northeast toward Urandangi, once a booming cattle town and gateway to the Channel Country, but now reduced to a rustic pub. This section of track was dry but deeply rutted, and the going was slow until we reached the Georgina River and the "'Dangi Pub" where thirst drove us to stop awhile. To celebrate our arrival into Queensland I had a XXXX beer (Queensland brew) and Barbara and Ian a lemon squash (lemon flavored soft drink). We chatted with the couple who owned the pub, Lis and Niels Knudsen, and learned that they were from Denmark, finding their way via a stint in Mt. Isa to the Queensland bush. We commented about an antique church pew against a wall in the pub, and the publican informed us with a twinkle in his eye that he and his wife sat on it each Sunday morning - and had a beer! They told us about a nice campsite along a creek about 80 km from the pub, just before hitting the Diamantina Developmental Road that leads south towards Birdsville from Mt. Isa. It was a lovely place with scores of Little Corellas, Galahs and Budgerigar s cavorting among the River Red Gums, scolding at us as we grilled lamb chops over Mulga coals.

The next morning saw us off and rolling by the early hour of 10:00 AM. Coming soon to the sealed Diamantina Developmental Road, we headed south through rather uninteresting country of grassy plains and Mulga scrub, soon arriving at the deteriorating township of Dajarra, through which I had last passed with Tony Lee in 1966. I had remembered it as a thriving small town, a railhead to which cattle were formerly droved for rail transport to the coast. Transport trucks have now replaced train transport of cattle, and Dajarra appeared to be settling into the desert dust. Some fuel, a beer, and a couple of ice creams saw us on our way with the next stop Boulia. In contrast, Boulia was a clean, thriving little country town at a three-way, paved intersection, rare in that part of the world. Since this was the last town before the turnoff to Sandringham Station, we topped off our fuel at the modern roadhouse, bought a few supplies, including milk and bread, and continued on our way, anxious to reach Sandringham. Just out of Boulia, the pavement ceased and we were on dirt track again, soon entering Diamantina Shire in which Sandringham Station is located. The country was transformed almost immediately from rolling uninteresting grassy plains into the familiar red sand hills, clay pans and tree-lined creeks of the Channel Country, all with an appealing postcard quality. While we traveled in our comfortable four-wheel drive vehicle with plentiful fuel, water and supplies, we tried to imagine the difficult trek of Burke, Wills, Gray and King, as they trudged north and then south through this same area over a hundred years before (1861), on foot with dwindling supplies and exhausted Camels. Of the four, only King survived the return journey, Gray dying between Birdsville and Cooper Creek, and Burke and Wills perishing after reaching Cooper Creek near Innamincka and discovering that their backup party had abandoned a depot there, only hours before their arrival (see Moorehead 1985).

Our progress was rather slow as the track passed through several boggy places that had not been graded since the last rains. Waterholes were numerous and we saw occasional groups of Emus here and there, and the large crane-like Brolgas (*Grus rubicundus*), long-legged, long-necked birds standing 3-4 feet high, in groups of 2-3 individuals. In addition, road-killed Red Kangaroos were frequent, and we saw one road-killed Dingo. It was rapidly approaching 5 PM, our usual camp selection time, and I was uncertain how far we were to the turnoff to Sandringham, and how far in from the turnoff Sandringham was located, as my memory had dimmed during the 18 years since my last visit. Knowing also that it would be impolite to arrive at the station at tea (dinner) time, especially on a Sunday evening, we elected to camp at the next favorable site, where clay pans and sand hills met. This combination, I remembered, had been attractive as campsites for Aboriginal people, probably for centuries if not for thousands of years, as indicated by a scattering of rock artifacts and flakes of all sorts, denoting former intensive use. The sand hills provided sources of plant and animal foods, while the clay pans collected rainwater; nearby gibber produced flint and quartzite rock for tool making. During favorable seasons and/or periods, these areas yielded all of the necessities of life and were heavily used;

the people retreated to the proximity of permanent waterholes during drought. I had promised Barbara and Ian an opportunity to examine such areas, and I hoped they still existed as I had remembered they had decades before. This campsite proved no exception to my memory. There was evidence that the entire lowland area had been under water not many weeks before, and, while some of the clay pans were dry, others still held water. All of the clay pans held a rich array of lithic scatter. Ian commented with his typical 8-year-old zeal "This place is an archaeologist's dream", and was off examining and collecting handsful of treasures, including fragments of grindstones, an entire grindstone, hand grinders, flinty and quartzite scrapers, and knife blades. He was frustrated only by the wealth of artifacts, having to put one down so that he could carry a still better one. Barbara and I watched with parental pleasure and pride his display of the joy of discovery. We had explained, and Ian knew from his experience at Uluru with the woomera, that these were national antiquities. Yet we also knew that it could be harmful to dampen altogether a blossoming spirit of inquiry, and so we took the liberty of granting him permission to keep eight objects of his choice among the dozens he brought back to camp, clutched in his hands. These eight were increased to ten when he couldn't choose between the last few. "Fair enough", we thought, and I, in particular, reflected on the importance of prudent collecting in my own development as a scientist, and also being reminded of the importance of collecting in the early development of nearly every field-oriented scientist of whom I am aware, including the young Charles Darwin with his beetles. I am convinced that balances can be reached between collection, preservation and conservation, be they archaeological or biological materials, and that these balances can best be learned early through actual experiences.

While exploring the margins of clay pans, and while setting up camp, each of us independently found shed carapaces and limbs of my old friend, the Inland Crab, which I had not noticed previously north of Sandringham. This was in the flood plain of the Georgina River, so I reasoned that they must be distributed throughout the entire Channel Country, including the Georgina. That night we had one of the worst camp meals in living history, as we tried to choke down tinned braised steak and onions. The only things that saved the meal were the accompanying noodles, although I had made the mistake of topping my noodles with the steak and onion glop. The next morning after breakfast, Ian and Barbara went off exploring the clay pans while I caught up with my diary. Ian came back to camp to fetch a bucket saying that he and mom thought that they had found a crab hole. As I concentrated on my notes, I didn't think much about it until I heard a shout "Dad, come here. We've found a crab". I rushed over to where they were standing near the edge of a clay pan, and they pointed at two holes they had filled with water from a nearby source. Barb pointed to one and said that something had come up to the water-filled entrance and stuck out its legs, and then had retreated. We all watched as the water bounced up and down, and pretty soon some sharp little legs poked out and were withdrawn. I was reluctant to grab it sight unseen for fear that it might be a spider or some other noxious critter, and sent Ian to camp for a tablespoon, with which he promptly returned. We watched, and as the legs emerged again, I plunged the spoon into the entrance and popped out a nice male Inland Crab, about 1.25 in. across the carapace. I was very pleased to confirm the presence of living adult crabs there, and Barb and Ian felt a real sense of accomplishment in identifying a crab hole, and getting its occupant to emerge for us. The crab was photographed and placed back into its hole into which it quickly disappeared, likely wondering what had become of the usual perquisites that accompany rain, like food and mates.

We reluctantly left this truly beautiful and interesting campsite at about 11:00 AM, drove down the Diamantina Developmental Road (i.e., dirt track) toward Bedourie, and came to the Sandringham Station turnoff about 5 km south of where we had camped. A sign indicated that the homestead was 61 km in from the main road. Finally, I had made it back to this land I had first visited 26 years before. As we drove down the rutted road past gibber, sand hills, clay pans, and creek crossings, it all began to have a familiar appearance, bringing back a flood of memories. The "rat-catchers" had returned, but this time in a rather different version of three MacMillens, well equipped with spotlight and hoop net. We planned to stop short of the homestead for lunch, and had to back up to the opposite side of a creek when the homestead suddenly loomed up before us. After lunch we proceeded to the homestead where we were met by Shirley Schrader, whose husband Geoff manages the station. Shirley invited us in for a "cuppa" (a cup of tea or coffee), introduced us to 15-year-old daughter Christina, and explained that Geoff was out mustering with all the hands. She was vaguely aware that we might be coming, soon warmed up to us, and invited us to stay in an unoccupied room in one of two stockmen's bunkhouses. The other one was fully occupied by the hired hands out mustering, although we were welcome to use the toilet and shower in that one. One room of ours belonged to "the dark bloke" (we later learned that his name was Hudson), who was in hospital for surgery. Thus, we had the place effectively to ourselves, except that Hudson's very pregnant small dog, "Bitty", came with the deal. A thorough sweeping and cleaning soon made the bunkhouse reasonably comfortable, with our kitchen set up in the screened verandah. It had a small electric fridge and running cold water that were handy. There was also an additional unoccupied room, where we stored our extra belongings, and where our friend from Alice Springs, Pip Masters, could bunk after her arrival a couple of days hence.

By the time we had moved into the bunkhouse, we were hot, sweaty and dusty, and so headed for the warm-water bore for a dip. It was just as good as I remembered, with warm artesian water gushing out of a 6 in. pipe ceaselessly, and into a cement-lined tub plenty big for the three of us. We washed, wallowed and luxuriated for about an hour, with Barbara and me mostly soaking, while Ian chased Spangled Perch in the runoff stream. It was good, then, to put on clean clothes to match our clean bodies and hair, and to wander over to the homestead for "tea", to which we had been invited by Shirley. After the previous night's braised steak and onions, a real station dinner sounded inviting, and it would be a relief from camp cooking, much more of which was in store for us ahead. Tea was interesting and varied, consisting of potatoes with beef gravy, fried eggs, steamed cabbage and carrots, and pizza; this was followed by a choice of several desserts: rice pudding, egg custard, ice cream, jelly (jello), and fruit topping. We met the remainder of the family, less Geoff: daughter Carolyn (age 10), and their young governess, Natalie (age 17). Natalie served as companion and teacher to Carolyn and Christina, who were studying at the station by correspondence, and School of the Air from Mt. Isa. Tea was accompanied by television, via satellite dish, new additions since my last visit.

At about 8 PM we excused ourselves, with thanks, to go spotlighting for mammals along the main station track leading out to the Diamantina Developmental Road. Shirley had advised us that "the rats (Long-haired or Plague Rats) have been plaguing since February, but they're not as bad now", so it was not surprising that we began seeing them shortly beyond the station, just as in 1974. We continued to see Plague Rats for the first 20 km out from the station, through grassland, gibber and creek country, but they began to thin out as we approached the first sand hill. Ian was just about asleep when we saw our first Fawn Hopping-mouse at about 30 km from the station, hopping from near the track out across a clay pan. Its ghost-like white color contrasted against the orange-red clay pan in such a way that at first I thought it was a large moth. Barbara, next to the window with the spotlight in hand, immediately had it in the beam and said, "it's a mouse". We weren't really prepared for mouse chasing at that moment, as the net was in the back of the truck behind the heavy tailgate, and not readily available. Nevertheless, I peeled off the road with the vehicle in pursuit, attempting to keep the headlight beams on the mouse, while Barbara held the spotlight as best she could out the window, directing the beam at the mouse. By this time, Ian was fully awake and cheering, as we careened across the clay pans, trying to follow the tortuous turns and erratic leaps of this handsome critter. It finally slowed down, and I thought I might catch it by hand and so stopped the car, got out with a rapidly-dimming flashlight, while Barbara got out, tethered to her spotlight by a 10 m lead connected to the truck battery. I followed the mouse about slowly, as it seemed to move only as rapidly as its pursuer, but each time I approached within grabbing distance (ca. 1/2 m) it would leap beyond reach. Had I my trusty net, I could easily have had it several times. By then Ian had come out to help me, and we managed to turn the mouse back toward the vehicle and Barbara. This proved to be a mistake as it was then between us and blinding lights. It soon ducked into the shadows of a small shrub and disappeared, after leading us on a very merry chase, certainly deserving its freedom.

We were more elated at seeing this one Fawn Hopping-mouse than disappointed at not having caught it. It also convinced us of the necessity of slightly better organization, so we got the net out from behind the tailgate and drove back to the road and continued on, with me driving, Ian in the center, and Barb at the window holding both the net and the spotlight. This didn't work either, for we lost the next hopping-mouse that appeared, as the net and the spotlight were too much for one person to handle. By this time Ian was anxious to try the net, and so we reorganized with Ian on the outside, holding the net through the window, and Barbara in the middle, holding the spotlight over Ian's head and out the window. We soon saw another hopping-mouse that I pursued slowly with the vehicle until it stopped. Ian got out with the net, and Barbara held the wee beast transfixed in the spotlight. Neither we or the mouse were as frantic as the first one we had seen, however this one was adept at escaping from under the side of the net, once confined. Ian proved to be both agile and persistent, and the third time he dropped the net over the animal he pounced on it and held it through the net, preventing its further escape. What a rat-chaser, and only eight years old! He was gleeful with his success and ready to push on for more, after I transferred the mouse to a specimen jar for safekeeping and later measurement. During the various turning movements, I had become disoriented and didn't really know the location of the road. Ian intuitively knew where it was and Barbara confirmed his intuition by locating the Southern Cross, and we were off down the road for more mouse mustering.

By confining our search to the clay pans and road near the base of a sand hill, we saw several more hopping-mice in short order, and Ian adeptly netted two of them, after a brief chase and some rapid footwork. We were all elated, and I was quite happy with three captures the first night, proposing that we head for the homestead and bed. Shortly after putting the net away behind the tailgate, we saw another hopping-mouse along the road, after which we gave cautious chase. It froze in the shadows of a small depression. Ian got out of the vehicle, crept up on it, and caught it bare-handed, winning the mouse-mustering championship of the evening with four captures. Shortly after, a much smaller mouse was seen behind a small shrub at the side of the road. This time I got out, crept up on it, and scooped it up in my hand, along with a painful handful of bindies (thorned seeds). It proved to be a lovely little mouse I had never seen before called Forrest's Mouse (*Leggadina forresti*), resembling very much some of our North American desert White-footed Mice (*Peromyscus* spp.). With that, we drove on back to the station, seeing only abundant Long-haired Rats along the way. That evening we spotted 51 Long-haired Rats, 9 Fawn Hopping-mice (capturing 4), and the 1 Forrest's Mouse. No marsupials of any kind were seen, even though much of the area traversed had yielded quite a few Kultarr (*Antechinomys laniger*) in 1966 and Kowari (*Dasyuroides byrnei*) in both 1966 and 1974.

The next two nights were spent spotlighting, too, but this time we were even better organized, having been joined by Pip Masters who had come from Alice Springs across the Plenty Highway and then down the Donahue Highway to Boulia, meeting us at Sandringham the second evening. On the second night of spotlighting, we drove directly to the clay pan country out 30 km from the homestead, where we had seen all of the Fawn Hopping-mice the previous night, noting numerous Long-haired Rats along the way. We also were alarmed to see three feral domestic cats crossing the road too far removed from the homestead to be homestead pets. At 30 km we stopped, Barb volunteered to spotlight from the roof rack, Ian insisted upon netting, and Pip rode shotgun from center seat. This proved to be an ideal combination as Barb could hold an animal in the spotlight, and Pip and Ian could approach it from either side (fig. 14).

In this manner Ian handily netted two, turning the net over to Pip, who also netted two, so we captured four of the eight hopping-mice we saw. We also saw, much to my alarm, two or three European foxes (*Vulpes vulpes*) that appeared to be stalking hopping-mice in the clay pans. These, together with feral domestic cats, are very substantial threats to small wildlife, and it was clear that both were now established in this remote sector of the outback. Our spotlighting for the night ended abruptly with the capture of the fourth hopping-mouse, and with Ian tripping in the shadows as he chased it toward Pip's net. He landed in a puddle of sticky clay ooze, soaking his boots, pants, and shirt. It was a cool night and, after removing his wet muddy clothing, Barb cuddled him warmly until we arrived back at the homestead close to midnight.

Figure 14. A pencil sketch by Ian MacMillen, age 8, of our mouse mustering team in action on Sandringham Station, 1992.

On the third night of spotlighting, we elected to try the track that went in the opposite direction, to the west of the homestead. The homestead grounds abut against a sand hill over which the track passes. We had scouted out the area previously by day, and knew that there was good clay pan country out several kilometers from the homestead. Our crack mouse-mustering team was unprepared though for the first hopping-mouse that emerged on a large expanse of bare clay near a cattle yard (= corral) just across the sand hill from the homestead. We sprang into action, Pip with the spotlight, Ian and Barb with the net, and after several miscues managed to bag it. We then reverted back to our successful arrangement of the previous night with Barb spotlighting from the roof rack, and Ian and Pip alternating with the net, and drove on, through a mixture of clay pans and gibber between sand hills, for about 15 km. Hopping-mice were sparsely present here and there, and we caught two more of five we saw, before turning back, cold and tired. In addition, Barb spotted two tiny mammals near a clump of plants along side the road, and we directed Ian to drop the net over the clump into which they disappeared. A careful search through the net revealed a single small insectivorous marsupial mouse, of the genus *Sminthopsis*, the other having disappeared (see fig. 15). We are uncertain of the exact identity but it likely was either the Striped-faced Dunnart (*S. macroura*), or the Fat-tailed Dunnart (*S. crassicaudata*), both of which have been observed at Sandringham before (Morton 1982a).

Figure 15. A dunnart from Sandringham Station.

One of the hopping-mice that eluded our net led us on the longest chase ever, first being spied near the road, then hopping over about 100 m of gibber to a series of clay pans at the edge of a sand hill. There, with Ian and Pip chasing, it would dart up the edge of the sand hill into the dense vegetation, but then would emerge back onto the clay pan, as if the sand hill was off-limits. It did this at least three times, being hotly pursued all the time; each time, though, abandoning the seeming safety of the sand hill for the more exposed clay pan habitat. The rat-catchers were finally bested by the rats, which eluded us all by disappearing into the shadows at the edge of a clay pan, leaving Ian and Pip exhausted and frustrated in their efforts. The behavior of this individual of seemingly preferring the clay pan habitat to the sand hill conforms precisely to the observations and conclusions Tony Lee and I had made more than two decades earlier on the habitat preference of the Fawn Hopping-mouse, *Notomys cervinus* (MacMillen and Lee 1969). That its preferred habitat is clay pans is confirmed further by our observations this time that all individuals of that species were observed in the immediate vicinity of clay pans, and nowhere else on the station. Another important, but alarming, observation that was made on this third night of spotlighting, was the sighting of another European fox, stalking hopping-mice in the clay pans, this time on the west side of the homestead. This was quite convincing evidence that foxes had become well established at Sandringham, along with feral cats, imposing a serious threat to small native animals in that region.

As we headed back to the homestead, we kept an eye open for more hopping-mice and, as the cattle yard behind the station sand hill came into view, we had just about given up hope of seeing any more. Just as we approached the expansive clay area near the yard Barb's spotlight caught a mob of three mice, each heading in different directions. She held the light on one that hopped to the right of the vehicle, Ian and the net dashed after another to the left, and soon all the mice were out of light-range and had disappeared. Shortly after, two more were spotted near the edge of the sand hill, and they employed the same strategy, darting off in different directions, and disappearing. Our organizational level was reduced to nil from too many late nights, exhaustion, and cold. Therefore, we decided to call it quits, and drive the short distance back to the homestead and to bed. As we approached the track over the sand hill, another hopping-mouse appeared, hopping a short distance along in the headlight beams, stopping as I slowed and stopped the vehicle. I was holding the net out of the window while driving, so that my companions could have their window closed to avoid the cold. I engaged the hand brake, put the Toyota in neutral, got out and slowly circled around beside the mouse, which appeared frozen in position. Fully expecting it to leap wildly away as had all the others, we were all surprised when I just walked up and dropped the net over it, without its moving. This was number four, equaling our captures of the two previous nights, and giving us a total of 12 for measurement and assessment of reproductive state. Barb remained on the roof rack and spotted one more hopping-mouse on our short drive back to the homestead. She said we almost ran over it as it was hopping along in the same track as our tires, toward us. It is also interesting to note that the first two nights we had traveled 30 km to the east of the homestead before seeing any hopping-mice while, on this third night, seven of the 13 hopping-mice were seen just over the sand hill from the homestead, not more than 150 m away. Thus proximity to the homestead and all of its activities, cats and dogs, has not resulted in local extinction of the Fawn Hopping-mouse, as I might have expected, even though the clay pan habitat near the cattle yard appeared to be highly disturbed.

In addition to spotlighting for small mammals, we did a small amount of live trapping around Sandringham Station, but caught only Long-haired Rats. Fifty traps set on the gibber east of the homestead, in an area where

I had both observed and successfully trapped Kowari (*Dasyuroides byrnei*) in 1974, captured only Long-haired Rats, 21 individuals. Fifty traps set by Pip Masters on sand hills 19 and 35 km west of the homestead on two different nights also yielded only Long-haired Rats, 19 animals the first night and 35 animals the second. These results, though, do not preclude the existence of other small mammals, which may simply have been overwhelmed by the abundance of Long-haired Rats. I am quite convinced though that at that time, for whatever reason, the populations of Kowari were very low, for we observed none while spotlighting, even in habitat where they had previously (1966 & 1974) been abundantly observed. The only possible Kowari seen was early one night (1930 hr), when Pip and Ian thought they saw one cross the road while driving across the horse paddock, about 5 km east of the homestead. Based on these rather casual observations it appears likely to me that during the ensuing years since 1966, the Kultarr (*Antechinomys laniger*) has become locally extinct from its old haunts on Sandringham, and will soon be followed by its dasyurid relative, the Kowari. I believe that the increases of feral house cats and European foxes (exotic predators) have contributed to the disappearances of these small native marsupials.

A summary of the small mammals we observed while spotlighting and trapping on Sandringham during this 1992 visit, and on previous visits, appears in Table 2. Certainly, in 1992, the Long-haired Rat was present in near-plague proportions, and the Fawn Hopping-mouse was present in the clay pan habitat in reasonably healthy abundance, although not as abundant as on previous visits. It is worth noting that the Fawn Hopping-mice were reproductively very active in June 1992, winter when days are shortest. Of the 12 individuals captured, five were male and seven female, and four of the females were conspicuously pregnant. Indeed three of them gave birth to three young each during their period in captivity. At least two of the five males had conspicuously enlarged testes, suggesting reproductive readiness. It was surprising to me to find them breeding before the winter solstice, as I am accustomed to the spring breeding of northern hemisphere desert rodents. Certainly at Sandringham, conditions were ideal for the successful production of young, as the recent rains were bringing on a flush of vegetative growth, promising an abundance of food. Observations by others (Breed 1982; Watts 1995) of reproductive periodicity in both the Fawn Hopping-mouse and its coexisting cousin the Spinifex Hopping-mouse confirm my suspicion that they are opportunistic breeders, able to come quickly into breeding condition following periods of rain and renewed abundance of food, regardless of season. Certainly, the Long-haired Rats had taken notice of these conditions, too, and had reproduced vigorously.

Table 2. Mammals live-trapped or netted on Sandringham Station, SW Queensland on three occasions between 1966 and 1992, using approximately equal capture effort

Taxon	Common name (*Species*)	Oct. 1966	Oct. 1974	June 1992
	Dunnart (*Sminthopsis* sp.)	1	1	1
	Kowari (*Dasyuroides byrnei*)	8	13	0
	Kultarr (*Antechinomys laniger*)	18	0	0
Rodents				
	Fawn Hopping-mouse (*Notomys cervinus*)	6	15	12
	Spinifex Hopping-mouse (*Notomys alexis*)	1	0	0
	Sandy Inland Mouse (*Pseudomys hermannsburgensis*)	0	1	0
	Forrest's Mouse (*Leggadina forresti*)	0	0	1
	Long-haired Rat (*Rattus villosissimus*)	0	40	40
	House Mouse (*Mus musculus*)*	0	5	0
Marsupial Species		3	2	1
Rodent Species		2	4	3

*Introduced

Our days at Sandringham were spent exploring the countryside. On one day, Shirley Schrader invited us to drive out to the mustering camp with her for a look-see, and to meet Geoff, which we did. Geoff and his hands were mustering all of the cattle on the station for Tuberculosis testing, because the adjacent station, Kamaran Downs, had had a "reactor", which we learned was an animal that tested positive for Tuberculosis. This had been a TB-free area, not requiring routine testing before sales, but a positive test in the region necessitated skin-testing the entire herd of several thousand animals, not a casual task. Mustering in the past had been conducted entirely on horseback. This was followed by a combination of horses and motorcycles, but mustering in 1992 was greatly simplified by use of a helicopter, both to spot stray herds and to drive them to a central locality, where they then could be "yarded" by men on horseback, for branding and skin-testing the lot. A veterinarian was present to confirm the test results, and no tuberculosis infection was observed. An awful lot of work just because a neighboring station brought in some infected cattle from outside the area. Geoff was seemingly taciturn about the entire matter.

One afternoon at the homestead Shirley notified us that Geoff and his stockmen were returning for the evening, and that we were all invited to a barbecue, after which we could stay and watch the "State of Origin" Rugby League match between arch rivals New South Wales and Queensland. We were pleased to learn that there was a sensitive side to this iron man of the outback, Geoff Schrader. Shirley also advised us that we might want to have a "bogey" in the stockmen's quarters before they returned. I thought that I had learned all of the outback vernacular, but a "bogey" was new to me, and our Australian-born translator, Pip, had never heard of one either. By context, we knew it had to do with personal hygiene, but of which part we were uncertain. At any rate, we took care of all possible needs before the stockmen returned, and probably had a "bogey" or two in the process. I later was to learn from Pee Wee and Diane Clarke, now at nearby Glengyle Station that "having a bogey" is simply washing up before a meal. The barbecue was short, but pleasant, and was held outside on the patio. The men sat on one side and the ladies on the other (I chose a safe, in-between spot) as we ate steak sandwiches and ribs. There was little conversation and what there was, was largely restricted to the ladies' side. Geoff did tell one funny story though. Apparently, some of the boys at one of the mustering camps had seen our vehicle out spotlighting, with lights flashing all around a clay pan in pursuit of a mouse. Geoff knew we were out spotlighting, but none of his hired hands was aware of our activities. When they asked Geoff what those lights were, he exclaimed "Min-Min" lights, and their eyes grew big. "Min-Min" lights are strange, luminescent apparitions that are reported occasionally near Boulia and Bedourie, and are thought by the locals to be supernatural, if not from outer space. We were glad to have provided at least that amount of entertainment. Rather than to stay for the televised footy game, which Queensland managed to lose much to the chagrin of our hosts, we excused ourselves to create some more "Min-Min" lights.

Our nights in the bunkhouse were not particularly restful what with the hard cement floor, a thin-becoming-thinner, inadequate sponge mattress, roving House Mice, and the pregnant dog, Bitty. Most of the House Mice were live-trapped and removed during the first two nights but, once trapped, we either had to lie there and listen to them rattle in the traps, or get up and carry the traps outside; and of course, they weren't all captured at once. While spotlighting the second night Bitty had her first pup, greeting us with it when we returned exhausted from chasing hopping-mice. No sooner were we in the sack than the House Mice started to cavort outside and inside the traps. I finally removed them and our captive hopping-mice, which were thumping in their containers, to the outside, hoping then for some sleep. As we slipped off into a deep slumber Bitty got up and began roaming about, leaving the newborn pup that then began squealing. Barb got up, comforted the pup, lured Bitty back to it and then the whole episode was repeated. Finally, about 4 AM Bitty delivered a second pup, and all became quiet, allowing us a little much-needed sleep, before getting out early to check our traps on the gibber. While we were gone, Bitty attempted to deliver a third pup that had breeched. Upon return, we found that its hindquarters had emerged, its umbilicus had been severed, and it was stuck. Barb became a midwife and helped the now dead pup to be delivered. It was apparent that Bitty had been in this predicament for some time as she had drug the partially-delivered pup all about our quarters, including on our sleeping bags, the floor, and here and there, leaving a messy trail behind. This, among other things, convinced us that it was time to abandon the bunkhouse and "go bush" once again.

We had planned to spend a couple of days and nights at one of three permanent waterholes to the west of the homestead, as there were earlier reports of Mulgara (*Dasycercus cristicauda*; Woolley 1989) near one, of considerable interest to Pip. Mulgara are small dasyurid marsupials of the arid zone, closely related to the Kowari. In addition, I remembered the waterholes as very picturesque places, teeming with birdlife and remains of Aboriginal camps. They were situated next to sand hills, and when I was on Sandringham Station in 1974, Terry Dawson and I visited them while exploring the country - one of them was the source of the duck and pigeon dinner we prepared for PeeWee and Diane Clark. In 1974, we reached the waterholes quite readily, as they were strung out at about five-mile intervals along the main track leading west of the homestead, toward the Simpson Desert. A map I drew in my field notes identifies them as the Seven-mile, the Twelve-mile, and the Seventeen-mile waterholes, names used by the Clarkes at that time. However, our explorations along the main track in 1992 revealed none of them, even when we proceeded along it more than 20 miles. They were large and permanent, and couldn't have vanished, particularly with the recent plentiful rain. Then it dawned on me that the track had been re-routed, explaining my dilemma. Directions obtained initially from Shirley Schrader didn't help much either, because the now obscure dirt tracks leading to the waterholes were located by landmarks that I couldn't identify, like "opposite the track to the Montara bore". Finally, Shirley drove Pip out to the turnoff to the Seven-mile waterhole, where we could then follow the old track on to the others.

We packed up our belongings, stopped by the homestead to say goodbye to Shirley, who presented us with a tenderloin of beef, and then proceeded behind Pip's vehicle to the Seven-mile waterhole for lunch. It was a lovely, big waterhole as I had remembered it, surrounded by Coolibah trees and with a red sand hill spilling down to it on one side. Birds were everywhere in the trees and air, particularly various parrots and raptors. After lunch I was anxious to press on to the Twelve-mile to set up camp for the next two or three days. It soon became apparent that the track that continued past the Seven-mile did not go by the waterhole, especially since we had by then proceeded well beyond the five-mile interval. After backtracking, we returned to the main road determined to follow Shirley's previous directions to the Twelve-mile. With fortitude, imagination and sheer luck we finally found it, helped by having seen the Seven-mile and what proved to be characteristic waterhole vegetation, particularly the Coolibah trees. We drove up to the water's edge on a grassy flat under several Coolibah trees, an ideal campsite. We soon observed that the two trees we chose to camp under each held a nesting pair of lovely Letter-winged Kites (*Elanus scriptus*), who viewed us nervously. In deference to them, we withdrew somewhat before setting up tents, tables, and other camp gear. After confinement to the bunkhouse and its fauna, we were relieved to be out in the open again, meandering.

Among the wonders of the waterhole, in addition to the kites, were myriad soaring Brown Falcons (*Falco berigora*), Straw-necked Ibis (*Threskiornis spinicollis*), several Brolgas (*Grus rubicanda*), Spoonbills (*Platalea* spp.), and quite a few Pink-eared Ducks (*Malacorhynchus membranaceus*). Here we were, on the western edge of the Simpson Desert, and there was permanent water that was recharged periodically by flooding, and teeming with water birds. Our amazement was increased further when Ian, while digging in the mud at the edge of the water, came up with a freshwater clam, and I caught a 6 inch Spangled Perch with what I thought was going to be a futile fishing attempt. Wonders never cease in the Centre! It was wonders related to this that first prompted Sidney Kidman in 1903 to purchase Sandringham and nearby Glengyle Station, to add to his chain of stations along the Georgina and Diamantina Rivers and their floodplains. He was aware that much of this country was periodically flooded, recharging both waterholes and lowland areas. These areas would carry rich grazing fodder for long periods after floods. Not only would these regions support large numbers of cattle, but they would serve, under a single ownership, as links in a chain of stations for overland droving of cattle to distant railheads and markets, greatly reducing the risk of inadequate water and food for the cattle on their long trek to market. Based on his own extensive observations of the Channel Country, Kidman knew that somewhere within the region there would be green islands of refuge for his cattle, even during extensive periods of drought. By 1912, near the peak of his career, Kidman's holdings of pastoral lands in central Australia constituted about 80,000 square miles; at the time of his death in 1935, Kidman's holdings included 68 stations, exceeding 100,000 square miles (Bowen 1991). This area is larger than the American state of Wyoming.

Although winter, the weather at the waterhole was mild, if not warm, and our down sleeping bags were uncomfortably warm. The beef tenderloin was put to good use with half of it grilled over Mulga coals the first evening, and the remainder saved for an excellent curry conjured up by Pip in a camp (Dutch) oven the next night. The nearby sand hills were covered with grasses and flowering shrubs to add to the splendor. Particularly striking were the Parrot Peas (*Crotalaria cunninghamii*) with their large, pea-like, yellow-green flowers. The sand hills were also littered with rodent burrows, which doubtlessly belonged to the ephemeral but now ubiquitous Long-haired Rat, as 50 traps yielded 19 of these critters one night, and nothing else. Ian was delighted to find here, too, as was predictable, an abundant scattering of Aboriginal stone tools and chippings, both on and between the sand hills. With the warm weather and abundant water came the nemesis of the outback, the bushfly. One day they finally drove Ian to retreat to the cab of the truck to read in relative peace, with the warm cab less threatening than the incessant insect.

Clouds and a lovely sunset accompanied our last night at the waterhole, and I was apprehensive about rain and being bogged in this remote spot, but the threatening part of the storm blew over by morning, allowing us to pack up in cooler comfort, and fewer flies. As Pip wanted to trap at the earlier-reported Mulgara site at the Seventeen-mile waterhole, we set off for it (we hoped) along the old track which Shirley had informed us would intersect the new, main track. Indeed it did, and at that intersection we could see about 100 m away at the end of a large sand hill the tell-tale Coolibah trees, denoting the probable presence of water. Sure enough, we followed an obscure track along the base of the sand hill to another large waterhole, the Seventeen-mile (also known as the Montara waterhole at the foot of the Montara sand hill). Twice, earlier that week, we had driven by this place on the main track, and had failed to notice the waterhole. We stopped and explored a bit, noting as we first approached the waterhole a pair of Brolgas with a downy young one; the parents flew away, and I just saw out of the corner of my eye the downy chick turn sharply at a right angle, approach a shrub and vanish. We stopped the vehicles and I explained what I had seen, pointing to the shrub. We approached and there, lying partially concealed by the shrub, was the downy chick, remaining absolutely still even when we approached to photograph it. Without doubt, this instinctive escape behavior provided concealing protection from predators, and a neat phenomenon to witness. Among other interesting observations at the waterhole and sand hill were numerous crude Aboriginal stone tools, and Ian and Barbara found a midden of empty freshwater clamshells, likely the leavings of tasty tucker. I noted with regret the clear impressions of feral cat tracks in the dried mud near the waterhole, attesting to the probable ubiquity of this unnatural predator in this remote sector of the country. It was time for Barbara, Ian and me to leave Sandringham, so we said good-bye to Pip, who would remain at the waterhole a couple of nights, attempting to locate a colony of her study animal, the Mulgara, on the nearby sand hill. She then was to meet up with her Ph.D. supervisor, Chris Dickman from Sydney University, to accompany him and other students to their study site at nearby Ethabuka Station, just northwest of Sandringham. Pip reported later that she was unable to locate any Mulgara on Sandringham; Pat Woolley (1989) reports their presence at the Montara sand hill in February 1968, but found none there in 1971 and 1979. It seems likely that this dasyurid species had also become locally extinct.

We stopped at the homestead on the way out to convey our thanks and good-byes, and to notify the Schraders that Pip was remaining at the Seventeen-mile waterhole for the next couple of days. Geoff met us as we drove up, explaining that Shirley and the children had gone to Betoota for the Gymkhana for the weekend, and we had missed them. This was a trek of about 800 km (338 mi) round trip, all on dirt track. Entertainment is few and far between in the outback! While saying good-bye to Geoff, I indicated that we were heading south through Bedourie to Birdsville, via Glengyle Station where we hoped to call in on PeeWee and Diane Clarke on our way through. I asked if Geoff could make radio contact and inform them of our visit. Geoff told us that PeeWee had been injured by a horse and was recovering in hospital, which put a damper on my plans to see them again, and to introduce Barbara and Ian. However, we could still visit Birdsville, the site of the famous Birdsville Races held annually the first Saturday in September, and dating back to 1886. Today people arrive for the horse races from all over the country. In 1990, around 4,000 people were present, arriving by private vehicle, 21 buses and 290 light aircraft. Luckily, we would be there in the off-season with a normal town population of 120.

As we drove out from the homestead through gates, past dry creeks lined with Gidgee Trees (*Acacia* sp.), over Gibber plain, and then finally through clay pans and sand hills, I began to reminisce about how this country had become so familiar to me, almost as if it was part of me. Without speaking, I knew that Barbara and Ian felt, intuitively, much the same way, although their exposures to it had been much briefer. Nostalgically I thought that this could well be my last visit to Sandringham, but I was elated that it would remain a memorable visit, certainly the highlight of our 1992 meanderings.

In Bedourie we fueled up and bought a few food items at the tiny store, which was new since my last visit in 1974 - the pub was the only functional business then. The storekeeper was pleasant and inquisitive and, on learning of our hopes to call in at Glengyle, notified us that PeeWee was out of the hospital, as he and Diane had recently come through town on their way back to the station from the hospital in Mt. Isa. Furthermore, PeeWee seemed fairly fit, and the storekeeper urged that we should by all means call in to see them. It was Saturday afternoon 6 June 1992, and it was apparent that our timing for arrival at Glengyle would be inappropriate, at about teatime, and that we had best camp for the night short of the station, delaying our visit until Sunday morning. After crossing King Creek (a southerly extension of the Georgina River, and named after John King, the sole survivor of the final exploring party of the Burke & Wills expedition) which was running with water, we turned off the road between two sand hills, and found a nice campsite in a dry, sandy wash. As we four-wheeled it up the wash Ian spied two large pieces of sandstone in the sand, so we let him out to inspect. He was elated to find two very nice Aboriginal grindstones, one quite flat and the other with a deep concave grinding surface. Obviously, others had liked this camping spot, too. A now dry Lignum swamp was adjacent to the camp, also, and in it we found shed carapaces of the Inland Crab. After dinner, while I secured the camp for possible rain that might ensue from the rapidly developing clouds and wind, Barb and Ian set 25 traps on the sand hill. Again, the storm blew itself out with only a couple of drops of rain, and we awoke to a clear, cool morning. The Long-haired Rat was obviously abundant here, too, as the 25 traps netted 19 animals, and nothing else. The productivity of this species is quite remarkable.

Not wanting to arrive at Glengyle at an awkward hour on Sunday morning, we got on the road early enough to arrive at the homestead about 11:00 AM. Glengyle is a large station with several outbuildings and small houses, and it had changed considerably since I had last visited Marg Arthur there in 1974. We pulled up to what I hoped was the main homestead house. We went through the gate into a lovely garden with lawns, flowers, and citrus trees - the homestead is on King Creek, with permanent water. I felt a little peculiar arriving out of the blue to call on people I hadn't seen for 18 years. Diane had heard us drive up and came around from the back of the house with a quizzical expression on her face, confronting us strangers. I explained who we were and her expression went from bewilderment to vague recollection to welcome amazement, all in rapid sequence. Diane welcomed us warmly, and called excitedly to PeeWee, who was in the house. He came out and underwent the same sequence of expressions followed by genuine warmth and welcome as he remembered. It was really a wonderful feeling to be greeted so warmly, almost like relatives, after so many years.

PeeWee was moving rather gingerly, but clearly was not letting his injury get him down. I asked him how the injury had occurred and he said he had been roping calves when his horse became excited and reared up and back, falling on him. He ended up with five broken ribs, and a punctured lung, but was "nearly right" now. He said it wasn't nearly as bad as when Reg Arthur had fallen from his bulldozer, and it had proceeded to run over him! Station life is not a bed of roses, as harsh realities are still an everyday part of life.

As we reminisced in the garden, Ian began to examine the margins of the carefully laid walks that radiated through it. He excitedly showed me that they were lined with Aboriginal grindstones and other stone artifacts that PeeWee and his stockmen had found on Glengyle; particularly appealing to Ian was the occasional black basalt (diorite) ax head which, before the introduction of steel axes by Europeans, were prized possessions, handed through the generations. PeeWee noted with affectionate satisfaction Ian's interest in these items, and said to him "Wait a minute," as he turned and went into the house. He returned shortly thereafter, carrying something in his hand that he held out to Ian and said, "Here, this is for you." Ian could hardly believe his eyes (nor could Barb and I), for PeeWee was presenting him with a large basalt ax head, carefully ground and polished with both care and use, about five inches in length. He told Ian that he must have ridden over the place where he had found it at least a dozen times, and then the next time it was just lying there. Ian was aware of the priceless nature of this gift, and accepted it gratefully in his rather shy manner, while PeeWee beamed at the response.

We were invited to join them for Sunday tea (at noon on Sunday), which was a huge meal of roast beef, potatoes and gravy, corn and peas, and cauliflower, with two kinds of jelly (jello), tinned pears and whipped cream for sweets (dessert). Diane still cooked for the station hands, and explained that she had made a new rule at Glengyle that the hands would eat in a dining room separate from that of the Clarkes, because often tea-time was the only time that she and PeeWee had a chance to discuss matters, and that they deserved at least this amount of privacy. The kitchen and dining facilities were in a building separate from the homestead house, and meals were announced to the hands by clanging a metal gong. We were just finishing this enormous repast when one of the hands straggled into the kitchen, looking blurry-eyed. Diane told him that he was too late to eat, but that she would save his plate for him for evening supper. As he glumly wandered back off toward the bunkhouse she explained that she kept very strict eating schedules, especially on weekends, and that some of the hands, including this tardy one, had had a very late night at the Bedourie Pub; "If they want to sleep in past noon, they can just miss their meals." PeeWee chuckled, clearly supporting this policy.

During the meal Diane told us proudly about her children: Ricky, the oldest, was now a stockman at a station near Tennant Creek in the Territory; LeeAnne, the second, was recently married, is a school teacher near Toowoomba, Queensland, and is an accomplished country music singer - Diane explained that "she accumulates degrees, rather than kids!"; Desmond, the youngest is a mechanic, and an Aussie Rules footballer, playing on a professional junior team in Queensland. Clearly, they were very proud of their children, and their varied accomplishments. While eating, PeeWee engaged Ian as often as possible in conversation, referring to him affectionately as "luv". He explained that he loved to hear his accent, and it was clear that he had a soft spot in his heart for children. We also discussed a little about Aboriginal workers on stations, and PeeWee explained that their former nanny-gardener Clara Reece, the rouseabout (handy-man) Hudson, and another had accompanied them to Glengyle from Sandringham. However, Hudson just couldn't settle down at Glengyle, explaining that Sandringham was his home ground, that he could only feel comfortable there; the Clarkes agreed that he should return - it was Hudson, whose quarters and dog Bitty we shared, who was in the hospital while we were at Sandringham. Upon asking about Clara Reece whom I had admired eighteen years earlier at Sandringham, we were informed that she had remained with the Clarkes at Glengyle until reaching retirement age, and then moved to Birdsville, where she died. Diane explained further that they and Clara wanted her to retire to Glengyle, but the government insisted that she could only receive her pension if she lived at the Aboriginal community in Birdsville. We learned that at Birdsville, Clara soon took to drinking, and within six months died as an alcoholic. This is a sad commentary on the government inequities that persist. Rather than living a comfortable, caring life for her remaining years at Glengyle, Clara's life was snuffed out prematurely and tragically in Birdsville, because of government policy. About the only distinction Clara will have is the brief statement in the Birdsville, 1991 Guide (Anonymous, 1991): "The most recent burial at the Birdsville cemetery was that of Clara Reece (Inchineeka) aged 68, on the 9th May 1982".

PeeWee also clarified a long-standing mystery that had haunted me since Tony Lee and I first filled up our petrol tank at nearby Betoota in

1966. The mystery was the extraordinarily long hose that Simon, the publican, used to deliver petrol from a large drum to a nearby vehicle; the hose extended beyond the pump a good 12 feet to the nozzle, far longer than was necessary. PeeWee explained with a grin: "Simon fills your tank, takes your money, and then sees you on your way with a wave and a smile. As soon as you are out of sight, he then holds the nozzle in the air and drains the liter or so of petrol back into the tank to be sold again. He must have sold that same hoseful at least a hundred times". Apparently, everyone else knew about Simon's trick but us, and put up with it cheerfully, as did we retrospectively, with a chuckle.

It was time to leave PeeWee and Diane, and Glengyle, so after a few photographs we said our good-byes, with sincere regrets in leaving the comfort of their open friendliness in this remote part of the bush. We pushed on south towards Birdsville, stopping only once to examine the old ruins of Cacoory Homestead, built in 1877, but long abandoned. Cacoory (AKA Carcoory) was once also part of the Kidman cattle empire, acquired by Sidney Kidman even before Sandringham and Glengyle, in 1899 (Bowen 1991). Birdsville had changed but little since my last visit in 1974, with the addition of a small store and a large petrol roadhouse as the main changes, and the establishment of a motel attached to the old (1884) Birdsville Pub, rebuilt numerous times. As evening was progressing and we were seriously in need of thorough "bogeys", we elected to stay for the night in the motel, at the rather exorbitant price of Au$70.00 for the three of us. It was clean, quiet and comfortable, with a cleansing shower, and probably a decent investment given our state of dishevelment. Barbara and I were still not hungry after our enormous lunch at Glengyle, and the thought of dinner was unappealing, but Ian thought he might be able to "choke something down". Therefore, we repaired to the famous Birdsville Pub, to soothe our throats with Victoria Bitter (me) and lemon squashes (Barb and Ian), and where we hoped to find something for Ian to eat. As the town was nearly abandoned, it was a quiet Sunday evening in the pub, with only a couple of colorful locals. On the quick food menu was a Zapburger, a frozen bacon-cheese burger that could be zapped in the microwave - high tech in the outback! Ian settled for the Zapburger and chips, and ate them with great gusto, as we quaffed our drinks.

We retired early to bed, prepared a camp breakfast in the motel room, aided by a hot water jug and toaster, and were off by 9:30 AM, a new record for us. Before leaving Birdsville, we drove out to the town's large billabong, on the Diamantina River, for a quick look. As we drove up, we flushed a large mob (ca. 200 birds) of Flock Bronzewing Pigeons (*Phaps histrionica*); these birds used to occur in flocks of thousands early in this century, but now are reduced drastically in numbers and in distribution. These are handsome birds, slightly larger than domestic pigeons (Rock Doves), with bronze back and wings, and iridescent blue and purple wing patches. In addition, the usual arid-zone parrots and white pelicans were around the waterhole. Barbara spied a pair of Brolgas with a downy chick, and photographed a dead Coolibah tree that was festooned with more than 300 roosting Little Corellas (see cover photo). With difficulty, we lured Ian back into the truck from his aquatic explorations, and headed north up the dusty Diamantina Developmental Road, on our return trek to Alice Springs. After passing by Cacoory and Glengyle, our first stop was at the Bedourie Store, only to be told by a local resident that the storekeeper was at the warehouse, but would be back shortly. As we had plenty of fuel and supplies to last until the next town, Boulia, we elected to proceed. The next stop was for lunch, near some clay pans just north of the Sandringham turnoff. Shortly thereafter, we had just left the Diamantina Shire, leaving the last sand hill behind as we entered rolling plains country, when an enormous snake crossed the road in front of us. I remembered the large King Brown Snakes (*Pseudechis australis*) that Terry Dawson and I had seen not too far from this location back in 1974, and reckoned that this was another. As we stopped the truck and gingerly got out with cameras in hand, the seven-foot monster continued unconcernedly on its way. Barbara was on one side of it, and Ian and I on the other. King Browns are deadly venomous, so I urged Ian to stay behind me in case the snake was aggressive. It was not, and we took several photographs as it continued on its way, stopping only to flick its tongue at the occasional rodent hole. It was sleek and fat, no doubt stuffed to capacity with Long-haired Rats. It didn't look quite right to be a King Brown, but there was no point in risking a closer examination. Later inspection of the photographs and guidebooks proved it to be a healthy specimen of the Woma (*Aspidites ramsayi*), a large sand python, which we had seen in captivity but never in the wild. Both the King Brown and the Woma have similar distributions in central Australia, both get to be seven feet in length, and both are a coppery brown color; however, they differ in head shape. It was thrilling to all of us to see the Woma, and a nice farewell to the Diamantina Shire, that had provided us with so many exciting adventures and lasting memories.

As the afternoon progressed, we decided to find a place to camp short of Boulia, and pulled off the road just south of town in a mixed grove of gum trees and Mulga, on the bank of the Georgina River. After setting up camp, we tried fishing in a large billabong nearby, using grasshoppers for bait. The pool was full of Spangled Perch, and we managed to catch and release several before it was time to prepare dinner. It became dark as we finished dinner so I turned on the propane lantern to illuminate the camp. This attracted the attention of a vehicle passing along the main road about 100 m away. We anticipated an angry landowner or a law officer, although the area was not posted nor had we crossed a fence. The vehicle drove up to the camp with headlights and dual above-the-cab spotlights shining brightly as I walked over to intercept it. The lone occupant was a rather scruffy young character in orange coveralls and thongs, with a six-pack of Queensland XXXX beer in his lap, who said "G'day, I'm Danny the Pigger!" He explained that he and his dog were out hunting feral pigs for the market, and, since we camped on the paddock where the local butcher ran a few cattle, he had stopped to see if we were Aboriginals who had decided to kill a cow and have a feast. He seemed relieved to find that we were tourists. It soon became apparent that the six-pack in his lap was not the first that had fueled the evening pig hunt. Not knowing quite what to expect, we watched as he emerged from the cab of his truck. I quipped that I couldn't offer him a beer because I had just finished my last one, but he said "no worries, mate; have one of mine" and handed me a XXXX as he lowered himself into one of our camp chairs near the fire. As he sat down, we noticed that his thong-clad feet were covered with dried blood, from a slain pig we hoped. He proceeded to tell us a series of yarns, including what a good deal pig hunting is, as you can make $200 - $300 a night, and never have to report the income to the government to be taxed. We were uncertain whether the pigs were killed with a firearm, but I didn't ask for fear he would bring one out of his truck. In the meantime, Ian was trying to get a small fire started that he had prepared in his own fire pit, without much luck. Danny said "let me show you, mate," and had even less luck, after considerable huffing and puffing. Barbara and I relaxed a bit when it became apparent that Danny was more lonely than he was dangerous, and he volunteered that a new butcher's market had opened in Boulia, and that if we mentioned his name we would get steak so tender that it could be cut with a fork. Finally, Danny the Pigger pulled himself to his feet, announcing his departure, and got into his truck with the remnants of the six-pack, and was off. Barb and I looked at each other with bewilderment and relief, and were reminded of a song from our favorite Aussie bush balladeer, Ted Egan: "Talk about a mob of characters in the outback; you meet them as you are traveling through the bush." Danny the Pigger was certainly one of those characters.

The next morning we were on the road to Boulia by about 10 AM, and drove into this nice clean little town as the shops were opening. We filled up with diesel at the roadhouse, bought some supplies at the Min-Min Store, a couple of tourist items at the Outfitters Store, and then stopped last at the butchers. I ordered some rump steak to grill for dinner, mentioning that Danny the Pigger had recommended the shop. The butcher, who was an attractive young lady, regarded us quizzically, saying nothing, while she wrapped the meat. Luckily, that evening, we had knives to augment our forks for cutting the meat.

We had decided to follow Pip Master's advice and take the Donahue Highway that she had used a week earlier just north of Boulia, and then link up with the Plenty Highway at the Queensland-Northern Territory Border. The Donahue proved to be a very good, recently graded dirt road, generally better than the earlier route we had taken via Urandangi, and much more direct. We stopped for lunch at a lovely billabong on the Georgina River near Glenormistan Station. After lunch, we crossed the Georgina, heading west toward the border and leaving behind the exceedingly interesting Channel Country that had made such an impression on us.

We entered the Northern Territory and joined the Plenty Highway at about 4 PM, and proceeded at a fast clip until about 5:30 PM when we began looking for a line of trees that might indicate a dry creek and a nice spot to camp. None emerged until finally about 6 PM we saw a vague line of trees paralleling the road about 1 km away. As darkness was coming on, we decided to four-wheel it over to the trees to camp. There were trees but no creek, but we found an adequate campsite, which was to be our last for the trip, as Alice Springs was less than a day's drive away. We got set up just as it became dark and had a last, good camp dinner of grilled steak with noodles. Although tired we read the evening story around the campfire, looking up into the sky occasionally at the myriad stars, and at the Southern Cross glowingly standing guard over the antipodes.

Our last day on the road was Wednesday 10 June, which was rather anticlimactic as last days of expeditions usually are. The routine of driving was broken by a noon lunch stop along a dry creek, whose sand was rose-colored and beautiful. The track we took to the creek was posted with a sign that said garnet fossicking (to fossick is to search), so we guessed that the sand was made up largely of tiny garnet fragments. This tweaked our curiosities and, since we had made good time, we decided after lunch to follow the track to the fossicking grounds. About 8-10 km up a rather dubious track we came to an area that looked like a minefield, and two retired couples were screening through the rubble looking for garnets. We got out, scratched around for an hour, and came up with a couple of handsful of rough, non-gem quality garnets, some of which were nearly an inch across. It was time to head back to Alice, so we piled in and drove back to the Plenty Highway, which soon became sealed, with no more dirt and dust. In short order, we were on the Stuart Highway, and then south to Alice

Springs, where we arrived at about 5 PM, with the vivid memories of an unforgettable visit to Sandringham Station and the Channel Country. The trip was 13 days and 3,257 km (2,036 mi), largely on dirt tracks through lovely, remote country the likes of which can be seen only in central Australia and by meanderings in the bush with your mates, be they family and/or friends

SUGGESTED FURTHER READING:

Berra, T.M. 1998. Natural History of Australia. San Diego: Academic Press.

van Oosterzee, P., and R. Morrison. 1991. The Centre: the Natural History of Australia's Desert Region. Balgowlah: Reed Books.

Chapter 7

WILDLIFE OF THE CENTRE: AS NATURE INTENDED

"There's always a drought somewhere in Australia"
Attributed to Sir Sidney Kidman,
Channel Country Pastoralist

What all these animals do in waterless land is inexplicable. Dogs and larger birds have their water sources, known perhaps only to them. The rats, mice and flies, however, must know how to replace water with something else. This much was certain; here, where they lived in their thousands, there was no water.
Hermann Beckler, Physician and Botanist assigned initially to the Burke and Wills expedition to the Gulf of Carpentaria; observation of 15 March 1861. (Beckler 1993 - translation of original German manuscript)

No environmental challenges in central Australia are greater than drought and, its reciprocal, torrential rain. The native organisms that live there do so in spite of these two extremes, and, in fact, their existences seem geared to continuing adjustments to life in a bipolar environment. Because the periods of drought are far more extensive than those of rainfall, the prevailing climate is one of aridity, but aridity that can be truncated suddenly by a superabundance of water. In the Channel Country of the Lake Eyre Basin, monsoonal summer rains in the tropical/subtropical north of the Northern Territory and Queensland may be sufficient "to bring the rivers down". This results in massive flooding further south, where little or no rain may have fallen, with water occasionally flowing as far south as Lake Eyre in South Australia, and even filling it. These floods may result in massive inundation of the nearly featureless terrain, leaving only the tops of sand hills emergent, forming temporary islands of refuge for the terrestrial animals that can reach them (see Flooding of the Georgina River, fig. 16).

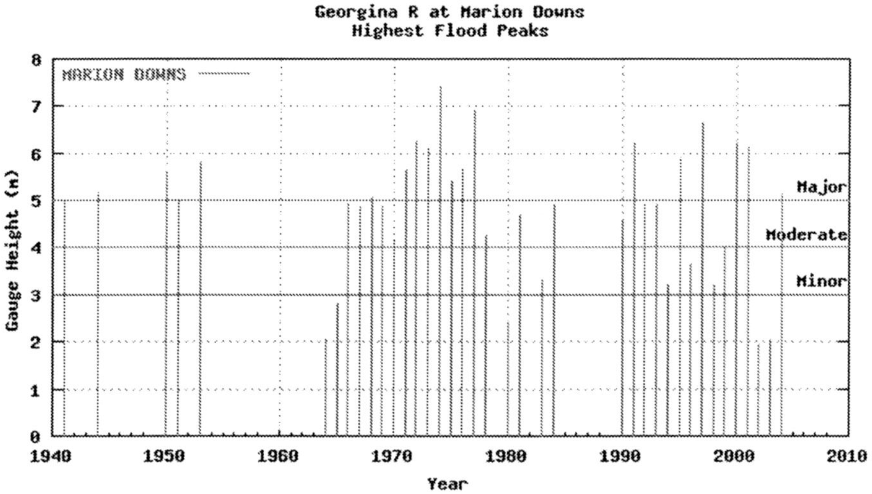

Figure 16. Graph of the unpredictable nature and intensity of flooding of the Georgina River over a sixty-year period as measured at Marion Downs, upriver from Sandringham Station. Reproduced from the online entry of the Bureau of Meteorology (Australia) Flood Warning System for the Georgina and Eyre Creek: Georgina River at Marion Downs - Highest Annual Flood Peak; copyright Commonwealth of Australia reproduced by permission (http://www.bom.gov.au/hydro/flood/qld/brochures/georgina/georgina.shtml).

A vivid description of such flooding, that occurred in March 1967 along the northwestern edge of the Lake Eyre Basin, is described by Alan Newsome (Newsome and Corbett 1975) following 6.5 years of drought: "The great rains resulted from straying tropical air and a southern cold front...The Finke River ran almost a kilometer wide at its highest, and flowed for six months that year, emptying itself into the Simpson Desert. Together with other streams, it filled clay pans and inundated swales between sand hills over great areas. To fly over the country at 10,000 m altitude was to see temporary ponds and lakes extending to the horizon on all sides all the way down to Lake Eyre. Water lay about for months, and the land had one of the flushest seasons in memory." For the aquatic and amphibious animals that inhabit this country, the occasional flooding provides an opportunity for individual and genetic dispersal and reproduction, followed by deaths of many when the waters recede and dry up, or survival of the lucky ones that can retreat to permanent waterholes or underground refugia. The recession of floodwaters is followed by a rejuvenation of pre-flood terrestrial habitats, with enhanced rates of annual and perennial vegetative growth and reproduction, followed rapidly by reproduction, repopulation and dispersal of the terrestrial animals that survived the floods. This relatively short period of plenty is nearly invariably followed by drought of unpredictable intensity and duration (see fig. 16).

However, even during periods of drought, some rain will fall sporadically somewhere in the Centre, forming temporary islands of vegetative production, separated by a sea of desert. Thus, the climate of the Centre can best be described as generally arid, but with rainfall that is unpredictable and sporadic in time, space, and intensity. The many ways that the native organisms respond to such an inhospitable environment, and yet seemingly thrive there, suggests a long history of exposure to alternating periods of drought and flood. According to J.M. Bowler (1982), arid conditions in central Australia date back to about 5 to 6 million years ago, followed by a period of oscillating wet-dry climates that lasted between about 2.5 and 1.2 million years ago. At least by one million years ago central Australia would have been arid, as it is today, followed by another oscillating series of wet-dry periods corresponding to the glacial-interglacial periods of the Pleistocene, and culminating in the interglacial period of today. Thus, the biota of central Australia has experienced several millions of years of exposure to conditions that have oscillated between wet and dry, but with increasing emphasis on the dry phase. During this time, the intervals of the wet and dry periods progressed from hundreds of thousands of years in the pre-Pleistocene Epoch to tens of thousands of years during the Pleistocene, and then to tens of years currently. Such a progression of climatic events would provide the kinds of natural selective pressures that would lead to the successful habitation and continuing existences of the very specialized flora and fauna characteristic of the Centre.

There is very convincing evidence that the Channel Country, the northerly portion of the Lake Eyre Basin (see map 3), experienced an intensification of these selective pressures during the latter part of the Pleistocene, and into recent times. These pressures could result in even further fine-tuning of the adaptive features of Channel Country biota. Recent studies indicate that the Lake Eyre Basin has been under the influence of the summer monsoons variously for at least 65,000 years (Johnson, et al. 1999). These studies are based on the carbon isotope composition of Emu eggshells from shoreline deposits of Lake Eyre, dating from present to 65,000 years ago. The investigators showed that the carbon isotope compositions of modern Emu eggshells reflect those of the plants they eat, and these in turn differed in regions dominated by summer monsoonal rains (central Australia) compared to those from regions dominated by winter rains (southern Australia). Comparing these modern eggshell controls with Lake Eyre fossil eggshells revealed that between 65,000 and 45,000 years ago the Lake Eyre region was under the influence of very effective summer monsoons. Between 28,000 and 15,000 years ago (the time of the last glacial maximum of the Pleistocene), carbon isotopes in eggshells suggest very weak summer monsoons; and between 10,000 and 1,000 years ago, the summer monsoons strengthened and were more erratic in occurrence. Thus the Emu and, we can infer similarly, the other plants and animals with which it coexisted, had at least 65,000 years in which to fine-tune their adaptive repertoires to life in the Channel Country.

As a person whose primary training was in Zoology, not Botany, I feel comfortable in discussing the basic adaptive features of the animals of the Channel Country, but not of the plants. The adjustments of plants to such extreme environmental conditions will be a fascinating tale to unravel and relate and will await revealing by plant experts, but I can at least make a start on the animal side of the story. For these animals, the region provides two general habitats, aquatic and terrestrial. The aquatic habitat may be either permanent, in the form of billabongs and waterholes in and around river courses, or ephemeral, as in playas and shallow pools that form temporarily during flooding. The far more extensive (both in time and space) terrestrial habitat has some features that are always permanent (sand hills), and others that are periodically, but unpredictably and variably, flooded (dry creek and river courses, clay pans, and Gibber plain).

For the aquatic animals the adaptive choices seem to be either to remain aquatic throughout all life stages and stay in a watery environment, or inhabit ephemeral aquatic habitats as long as they persist, but have a drought-resistant dormant life stage that can wait for months or years on, in, or under the dry surface until flooding once again occurs. They therefore may be subdivided into obligatory and facultative aquatic organisms, respectively. Among the obligatory aquatic forms I have seen in the Channel Country are several species of fishes, turtles, clams and mussels, shrimp and crayfish, and water birds. It is only the birds, including ducks, spoonbill and ibises, that can move by flying between water bodies during drought; they are more properly referred to as nomadic in life style, and will be considered in a later section that deals with nomadic animals. The more strictly obligatory aquatic animals can disperse only during flooding, and those that end up in temporary water bodies perish as these dry up. All permanent waterholes and billabongs that I have witnessed, no matter how remote, have possessed some of these aquatic animals, demonstrating the powerful dispersal capabilities of these species during flooding. Also, I have witnessed dry and drying clay pans following flooding that were littered with dead fish, attesting to the risky nature of this existence for individuals. Certainly, these conditions constitute very powerful natural selection agents for successful dispersal, but chance survival may also play an important role.

Among the aquatic animals that successfully inhabit ephemeral pools and ponds by having a drought-resistant dormant life stage (the facultative aquatic animals) are the Branchiopod Crustaceans, the Inland Crab (a decapod), and several species of frogs. The Branchiopod Crustaceans, including Fairy Shrimp, Clam Shrimp, and Shield Shrimp, that inhabit temporary ponds and pools, are capable of producing early developmental embryos called cysts, with thick, protective coats that are resistant to heat, cold, and desiccation (see Eriksen and Belk 1999). Adult females, finding themselves trapped in drying pools, release these cysts prior to dying, leaving behind the next generation as dormant embryos with arrested further development. These cysts will hatch and reinitiate development into adults only with the next inundation, whether that be months, years, or even decades later. At least the dormant eggs of the Shield Shrimp may be carried aloft from dry clay pans by wind into the air, to be deposited miles away, or on such remote places as the depressions atop Uluru, to develop into adults after rain, as we witnessed.

Equally remarkable are the frogs and the Inland Crab that survive periods of drought as adults without access to surface water, by retreating underground into burrows as surface waters recede, and then entering into states of suspended animation until the drought is over. Although about as distantly related as two kinds of animals can be, these frogs and crabs have converged in several of their adaptive solutions for survival in such a precarious environment. Both choose fine-grained clays that are tenacious of moisture in which to establish burrows, and retreat to these burrows during drought with substantial stores of energy-rich fat. Both reduce their metabolic rates during drought to about 30 per cent of that of surface-active animals, increasing the probability that their energy stores will last until rainfall again permits resumption of aquatic life and feeding. Here the similarities end, as each has evolved additional specializations that enhance survival during drought.

Frogs, such as *Cyclorana platycephala*, one of the water-holding frogs, drink copiously as their temporary ponds dry, and retreat to the burrow bottoms with urinary bladders filled with extremely dilute urine. These frogs have a highly vascularized belly region that is capable of taking up water from burrow surfaces until the soil becomes quite dry, replacing any water lost through the skin. As burrow surfaces continue to dry, this frog undergoes a partial skin molt that eventually forms a nearly impermeable cocoon around the entire body, with the exception of the nostrils that are open to burrow air through tubular extensions, permitting breathing. This cocoon reduces water loss through the exposed surfaces to about 3 per cent of that of frogs without cocoons. This greatly reduced rate of water loss in cocoon-bearing frogs is readily made up by absorbing water that is stored in the bladder as dilute urine. Van Beurden (1982), who conducted these studies, estimates that adult dormant frogs in burrows, with reduced rates of energy metabolism and water loss, have sufficient energy and water stores to withstand at least two to three years of drought. When the rains come to provide temporary surface water, the frogs emerge from dormancy, rapidly replacing depleted water and energy reserves (the latter by feeding), and enter a reproductive phase. Female frogs deposit eggs in the water that are simultaneously fertilized by males, and the eggs undergo accelerated development, as do the resulting tadpole larvae; these rapidly develop into small frogs, which feed and grow at accelerated rates, readying for the next drought cycle.

The Inland Crab, on the other hand, employs another suite of adaptive characteristics to withstand drought while in their protective burrows (see Chapter 3, and Greenaway 1984, for detailed explanations). The branchial chambers house both the gills for water breathing and lungs for air breathing. Their exoskeletons and lungs, together, exhibit greatly reduced rates of water loss, approaching rates more characteristic of reptiles. In the aquatic phase crabs excrete nitrogen in the form of readily soluble but toxic ammonia, but while in the burrows, nitrogen elimination ceases, and it is apparently stored in the body fluids as an as yet-to-be-determined non-toxic substance. Under extreme laboratory conditions, the crabs are tolerant of high levels of body water loss and resulting concentration of body fluids, with complete recovery following rehydration. Crabs excavated while dormant in burrows during drought had body water contents and concentrations characteristic of fully hydrated animals in the laboratory. These observations, and our intuition, suggest that the crabs possess some mechanism that allows them to remain hydrated within their burrows. We suspect that they are able to take advantage of the vertical temperature differentials that occur in their burrows over the 24-hour period: by crawling to the cooler burrow entrance at night, allowing the body temperature to equilibrate with the low night-time surface temperature, and then rapidly descending to the warmer, humid burrow bottoms where condensation of water vapor on the carapace would likely occur; our laboratory investigations have shown that such water droplets coalesce and flow toward the mouth region, where they are imbibed, resulting in a fully-hydrated state. Assuming that Inland Crabs in their burrows during drought are able to remain fully hydrated, or dehydrated within tolerance levels, then energy stores may be the factor that limits periods of underground dormancy during drought. Some preliminary data suggest that crabs enter burrows during drought with sufficient energy stores in the form of fat, gained during aquatic feeding, to sustain them for at least a year (Greenaway, personal communication). Their distribution throughout much of central-eastern Australia (including the Channel Country) would seem to confirm that they survive droughts while in their burrows for at least these durations. Although their reproductive habits have not been thoroughly investigated, we do know that females bear progeny while in the burrows, in anticipation of the summer monsoons. Eggs likely are held in the brood pouch, where they soon develop directly into crablets, without an aquatic larval stage; there they are nourished on egg yolk. If the drought breaks, the female becomes aquatic, releasing her crablets into the temporary waters to fend for themselves. If the drought is prolonged, we suspect that the female eventually eats her crablets, recycling the conserved nutrients to apply later to another reproductive attempt. This scenario presupposes that mating occurs during the aquatic phase, and that the females may store sperm for more than one reproductive cycle.

Thus, here are two kinds of truly remarkable creatures, frogs and crabs, that have successfully occupied Australia's arid Centre by spending probably in excess of 90 per cent of their adult lives in underground sanctuaries, waiting for the next rain!

Whereas the aquatic animals of the Channel Country may be subdivided into obligatory and facultative aquatics, as explained above, the terrestrial animals more readily fit into categories that are based upon relative mobility: nomadic or sedentary. The nomadic terrestrial animals that I have encountered include most birds, bats, large, flightless critters like emus and kangaroos, and at least some of the flying insects, all capable of moving fairly rapidly on or over land in search of food and water. However most of the animals of the Centre have only limited mobility, and are much more sedentary in their existences. These sedentary animals include myriad kinds of invertebrates (insects, spiders, scorpions, millipedes, centipedes, etc.), snakes, lizards, some birds, and small mammals (both marsupial and placental). Being sedentary and only capable of local movements, these animals must rely on the resources available in their immediate habitats, regardless of climatic variability.

Among the insects of the Australian arid zone, termites are probably the most heralded, largely because of the habit of some species of building large, conspicuous red-earthen mounds above the soil surface. However, I have not witnessed termite mounds in the Channel Country, and conclude that they are either absent there or rare, suggesting that the mound-building termites avoid these regions that are periodically flooded when the rivers come down; they are quite common further north. In contrast, harvester ants of several species occupy many habitats in the Channel Country. At Sandringham Station, five harvester ant species occur on a stony plain (gibber) with tussock grassland, eight species on a sand ridge with Canegrass, and nine species on sand plain with Spinifex (Morton 1982b). These ants that forage for seeds on the surface but shelter in fixed underground chambers as colonies, are potentially susceptible to the effects of periodic monsoonal rain and flooding. While there are many physical variables that distinguish Morton's (ibid.) ant study sites, it is interesting to note that the stony plain site is the one most likely to be inundated for extensive periods (Morton personal communication). This site has the fewest harvester ant species and the lowest ant species diversity. Further study of the relationship between harvester ant habitat choice and probability of inundation is warranted. Other sedentary arthropods that attracted my attention at Sandringham Station were burrowing spiders that build elaborate burrow openings that are resistant to flooding. One of these establishes a thick bathplug-like, hinged door that is watertight when closed, and is commonly found on floodplains and clay pans (Main 1982). Another spider species that particularly aroused my interest is one that builds a snorkel-like, impermeable turret that extends several inches above the soil surface, usually leaving the opening above flooding water levels (ibid.). Little is known of the life histories of these spiders, other than that they are sit-and-wait predators, pouncing upon other terrestrial arthropods that blunder into the vicinities of their burrow entrances; even during drought they likely derive ample food energy and water as long as prey are available.

Our visits to the Channel Country were always in spring or fall, purposely avoiding the extreme summer and winter seasons. Since these visits coincided with periods of moderate temperature that were often lower than optimal for surface activity of the ectothermic reptiles, we saw only a few lizards and snakes. The lizards included the very common Bearded Dragon (Family Agamidae, *Pogona vitticeps*), a handsome, foot-long lizard seen on warm days, sunning on rocks, tree limbs and stumps, and fence posts; Bluetongue and Shingleback Skinks (Family Scincidae, *Tiliqua multifasciata* and *Trachydosaurus rugosus*, respectively), also about a foot long, with heavy, cumbersome builds; and Spencer's Monitor (Family Varanidae, Varanus spenceri), a large (up to four feet), sleek, active predator on other lizards, birds and mammals. Although we witnessed only a few species of lizards in the Channel Country, our observations are likely not representative of their real abundance. Pianka (1986), in a very comprehensive field study of lizard faunas that inhabit the Great Victoria Desert of Western Australia, reports that the species diversity of lizards there is extremely high with as many as 15 to 40 species at a given site. Pianka's study site habitats, being in a region of summer monsoonal rains, were rather similar to those that we encountered in the Channel Country. Pianka (ibid.) found this abundance of lizard species was positively correlated with the unpredictability of annual rainfall which, in turn, probably yielded correspondingly unpredictable levels of annual primary and secondary production, which somehow influences the larger number of lizard species that may co-occur. Therefore, it is likely that there are far more lizard species in the extremely unpredictable Channel Country than our casual observations revealed. Although central Australia is often thought to be teeming with poisonous snakes, we have only encountered two kinds of living snakes, and on two occasions, in the Channel Country. In 1974 Terry Dawson and I observed two venomous King Brown (or Mulga) Snakes (Family Elapidae, *Pseudachis australis*), each between six and seven feet in length and crawling unconcernedly across the desert floor. In 1992 Barb, Ian and I came across a large (also in excess of six feet) Woma (Family Boidae, *Aspidites ramsayi*), giving it wide birth because of its close resemblance to the very venomous King Browns - we were only later able to identify it through photographs as a non-venomous Sand Boa. Although perhaps coincidental, both of these sightings occurred during population outbreaks of the Long-haired or Plague Rat (*Rattus villosissimus*), which would make a fine meal for these snakes.

To my knowledge no systematic collection of reptiles has occurred that encompasses all of the Channel Country, so their numbers of kinds and distributions remain uncertain. However, Tyler et al. (1990) have compiled a list of reptiles collected and/or likely to occur in the lower half of the Lake Eyre Basin that lies within South Australia. This list includes 85 lizard species and 24 species of snakes, most of which also are likely to occur further north in the remaining portion of the Channel Country that lies within Queensland and the Northern Territory. This list suggests, as does the work cited above of Pianka (1983), that an enormous species diversity of lizards exists in the region, and that an equally impressive array of adaptations must be present to enhance existences in this challenging environment.

Hints of some of these adaptations were gleaned from a field study Mike Augee (Zoologist at The University of New South Wales) and I conducted in a desert shrubland in northwestern New South Wales during 1974 and 1975 (see MacMillen, et al. 1989). We examined daytime temperature preferences, seasonal activity, and diet of four species of coexisting lizards near the University's Fowlers Gap Research Station, about 400 km SE of the Lake Eyre Basin (ibid.). The four species are the Central Bearded Dragon (*Pogona vitticeps*), the Netted Ground Dragon (*Ctenophorus nuchalis*), the Earless Dragon (*Tympanocryptis lineata*), and the Shingle-back Skink (*Trachydosaurus rugosus*), all of which are on Tyler, et al.'s (1990) list for the Lake Eyre Basin, and therefore are likely to occur further north in the Channel Country (where, indeed, I have seen both Bearded Dragons and Shingleback Skinks. They varied considerably in body mass, with the skink the largest (ca. 715 g), followed by Bearded Dragon adults (ca. 237 g), the Netted Ground Dragon (ca. 25 g), the Earless Dragon (ca. 5 g), and Bearded Dragon juveniles (ca. 3 g). We found that the largest species (Shingleback Skink) avoided surface activity altogether during hot midsummer days and the cold midwinter period, but was surface-active during cool mornings of warm days and during midday of cooler seasons. The less large Bearded Dragon was active during all seasons except the coldest winter period, but abandoned surface activity during the hottest midday portion of midsummer. The smallest species (Netted Ground Dragon and Earless Dragon) appeared to prefer the hottest summer days, and were surface-active throughout, but avoided cool and cold seasons. While the three larger species were predominantly herbivorous, the smallest species and the tiny juvenile Bearded Dragons were wholly herbivorous, wholly insectivorous, or dietarily mixed. Although the large Shingleback Skink was predominantly herbivorous, its diet was by far the most varied, even including carrion, and sheep and/or rabbit feces.

Thus, to extrapolate these findings to the Channel Country, it appears that the smallest lizard species and individuals confine their daytime activities to the hottest periods of the hottest season, which coincides with the summer monsoons (when they occur). The resulting availability of their preferred foods of insects and/or green vegetation provides both energy and water. During prolonged periods of drought, it is my guess that many of these small lizards perish of starvation and dehydration, but somehow others must survive to ensure subsequent generations. The large Shingleback Skink is a cool (not cold) temperature specialist with the most generalized diet and limited mobility, all of which combine to promote survival. It most likely is in underground retreats when monsoonal (hot weather) flooding occurs, and when it emerges in cooler weather it may subsist on the wide variety of foods its more limited locomotor capabilities enable it to encounter. The next-to-largest adult lizard (Bearded Dragon) is active under a broad range of moderate temperatures, avoiding surface activity only during the cold winters; its more limited herbivorous diet is consistent with relatively greater powers of mobility that are consequences of a lithe, agile build and large size that facilitate access to preferred edible plants. That this dragon is active under a fairly broad range of temperatures is consistent with the observations of Bartholomew and Tucker (1963) that a close Queensland relative, The Eastern Bearded Dragon, when placed in an environment of changing temperatures, has a body temperature that rises more rapidly than it falls. They conclude that this capacity is controlled by manipulations of metabolic rate and the cardiovascular system, and permits the maintenance of a fairly constant body temperature, somewhat independent of environmental temperature, resulting in more prolonged periods of activity in a fluctuating thermal environment. Little to nothing is known about how these and the many other Channel Country lizards, or snakes for that matter, deal with sustained drought or flooding, and how their reproductive periods respond to these climatic variables. Hopefully, additional meanderings by some enterprising biologist(s) will begin to fill these knowledge gaps.

While most of the birds that inhabit the Australian arid zone may be considered nomadic, Schodde (1982) cites the occurrence of members of six avian families that are sedentary with limited mobility, and unlikely to move to the widespread but patchy flushes of food and water that follow sporadic rain. These, he states (ibid.), are members of non-drinking families that rarely or never drink water. They are chiefly small fruit- or insect-eating birds, whose diets are quite succulent when available, but how they withstand prolonged periods of drought has yet to be determined. Schodde's statements are reinforced by direct observations of drinking behavior by arid zone birds at isolated waterholes during hot dry seasons at eleven sites scattered throughout central Australia, including the Lake Eyre Basin (Fisher, et al. 1972). These authors (ibid.) found that of 118 bird species observed, 49 were seen in the vicinity of water but were never seen drinking; these birds were all either fruit-eaters or insectivorous and/or carnivorous. Schodde (1982) also comments that these sedentary birds are strict spring breeders whose average life spans may be shorter than local drought, implying successful spring reproduction even in the absence of rain. Although the distributions of some of these sedentary species encompass portions of the Channel Country, none of these birds caught my eye, but they certainly deserve further attention.

Among the small mammals of the Channel Country, sedentary because of their size-induced more limited mobility, there are a few species of placental rodents (Family Muridae) and more species of insectivorous/carnivorous marsupials belonging to the Family Dasyuridae.

While the incidence of dasyurid species in the Channel Country is nowhere as great as that presumed for lizards, Steve Morton (1982a) documents that 45 per cent of all dasyurid species occur in the arid zone compared to 30 per cent of all other mammal species (bats excluded). This suggests that dasyurids represent an unusually diverse and successful mammalian fauna in the Australian arid zone. Morton (ibid.) cites the occurrence of seven dasyurid species on Sandringham Station, more than at three other arid zone sites that have been studied, and more than the five rodent species I have seen at Sandringham. Further, Morton (ibid.) implies that these dasyurids, whose carnivorous/insectivorous diets approximate those of many of their co-inhabitant lizards, may owe their relatively high species diversities to environmental influences shared with the more diverse lizard fauna: low, unpredictable rainfall with sporadic wet years. For both lizards and dasyurid marsupials of the arid zone the suspected links between climatic variability, high species diversity, food production, and reproductive periodicity have yet to be examined and disclosed, fruitful areas for future research.

As happens so often in Biology the laboratory performances of dasyurid marsupials are much better known than their field performances. In the laboratory, animals can be manipulated experimentally, using statistically adequate sample sizes, while making sophisticated measurements of responses to carefully controlled environmental conditions. In the field the investigator is at the mercy of a changing environment and the vagaries of the animals, making field studies logistically much more difficult and time consuming than laboratory studies. In the laboratory a complete study of thermoregulation and energetics of a small animal might be accomplished within a matter of weeks, while a satisfactory field study of the same animal might require a matter of months and frequently a year or more, while seldom providing the same degree of completion as the lab study. It is tempting to focus on laboratory studies, because of their relative ease, but it is the essential fieldwork that provides the questions that are pursued in the laboratory. Each of the seven dasyurid species that occurs on Sandringham Station has been the subject of detailed laboratory studies of physiological responses to simulated environmental conditions (Table 3).

These dasyurids, being insectivores/carnivores, are the ecological and visual equivalents of North American shrews and weasels. Since these seven encompass the full range of body mass of the dasyurids most frequently encountered in the Channel Country, I will use them to develop a composite picture of the functional attributes likely to contribute to their success there. In terms of energy metabolism dasyurids in general have

Table 3. Dasyurid Marsupial species observed on Sandringham Station, their body masses, and status of physiological investigation (after Morton 1982a, Geiser 2004).

Species	Body Mass (g)	Status
Planigale gilesi	10	Energy Metabolism
Planigale tenuirostris	7	Energy Metabolism
Sminthopsis crassicaudata	15	Energy & Water Metabolism
Sminthopsis macroura	20	Energy metabolism
Antechinomys laniger	30	Energy & Water Metabolism
Dasycercus cristicauda	100	Energy & Water Metabolism
Dasyuroides byrnei	110	Energy & Water Metabolism

basal metabolic rates that are reduced by 32 per cent of those of their placental mammalian counterparts (MacMillen and Nelson 1969); this applies when the animals are at rest in thermally neutral conditions (ambient temperature ca. 30-34°C), conditions likely encountered primarily during the summer while in their burrows by day. While this represents an energy savings (compared to placental mammals), its importance is probably negligible as it occurs when temperatures are mildest, prey is most abundant, and when caudal and body fat storage is likely to be at its maximum, providing an ample, renewable energy supply. Under conditions of lower temperatures and increased locomotor activity, as would be encountered on the surface at night, metabolic rates of dasyurids are indistinguishable from those of their placental counterparts (Hinds and MacMillen 1984; Baudinette 1982). The primary energy conserving mechanisms employed by dasyurids during times of low food supply and high energy demand (cooler/colder weather and possibly summer drought) appears to be a combination of mobilizing energy stored in the form of fat deposits and the use of torpor. By converting some of the energy gained from an arthropod diet during times of plenty into internal fat stores, an energy reserve may be acquired that can be drawn upon when surface food is in lesser supply or temporarily unobtainable. This fat store may then be gradually drawn upon while retreating to a protective burrow, and entering a state of torpor in which body temperature drops to near that of the burrow, accompanied by a concomitant decrease in metabolic rate that is considerably lower than when operating at a normal body temperature. During torpor, the metabolic rate ranges between 13 and 54 per cent of the basal metabolic rate (Geiser 2004). The extent to which torpor is employed by dasyurids in the field is unknown, except for *Sminthopsis crassicaudata* (aptly called the Fat-tailed Dunnart!), which uses torpor infrequently and only during the colder part of the year when foraging success was minimal (Morton 1978). Thus far, torpor has been demonstrated in all of the dasyurids species observed on Sandringham Station, and, at least in the laboratory, torpor appears to be restricted to daily rather than long-term (hibernation/aestivation) use (Geiser 2004). Further, this daily torpor is used effectively as an energy-conserving device even when pregnant or lactating, without retarding development or growth of the progeny (ibid.). Thus the combination of torpor and fat storage serves these animals well, apparently ensuring the survival during drought and/or cold winters of at least a sufficient number of individuals to yield successive generations until conditions improve.

Most small sedentary desert animals gain energy and water from the same food packet, which, in the case of dasyurids, is arthropods and/or small vertebrates whose bodies are about 65 per cent water. Laboratory studies of the Mulgara (*Dasycercus cristicauda*) and the Fat-tailed Dunnart (*Sminthopsis crassicaudata*) reveal that they can maintain weight and positive water balance on an animal diet without drinking water, while not subjected to heat stress (Schmidt-Nielsen and Newsome 1962; Morton 1980). It is likely that dasyurids seldom experience heat stress, as they are active on the surface only at night, spending the day in thermally buffered shelters. Schmidt-Nielsen and Newsome (ibid.) found further that the Mulgara produced highly concentrated urine with large amounts of urea from the nitrogen-rich diet, thereby reducing urinary water loss to a minimum. Studies of the Kowari (*Dasyuroides byrnei*) from Sandringham Station by Terry Dawson and me (1986) attempted to provide an additional dimension of insight into the water economies of dasyurids by examining effects of locomotion at different temperatures on water loss by evaporation and water gain (both preformed and metabolic) from an insect diet. Evaporation accounts for nearly all of the water lost while an animal is engaged in active locomotion while foraging. We (ibid.) simulated foraging locomotion by running animals on a treadmill at several speeds and temperatures, while measuring simultaneously evaporative water loss and oxygen consumption; the rate of oxygen consumption could be translated into metabolic water gain, and preformed water gain could be determined from the diet of juicy crickets. We found, at all temperatures (15, 25, 33°C) and running speeds (1, 2, 3 km/hr) tested, that water gain surpassed evaporative water loss, and that this differential increased with increases in running speed and decreases in temperature. Our conclusion was that this dasyurid remains in positive water balance while foraging on the surface at night, assuming that food is found, adding to our earlier observation that the same animals, while caged and at rest, maintain positive water balance on an animal diet.

All of these laboratory studies combined argue very strongly that the dasyurids of the Channel Country need never drink water as long as they can find their preferred food of arthropods and/or small vertebrates, providing them with all the water they need. Further, this same food provides all of the energy that is required for maintenance, activity, growth, and reproduction. In times of food scarcity resulting from prolonged cold periods and/or drought, it is likely that only the more successful foragers find a dwindling food supply, as both predator and prey experience diminishing populations, only to rebuild when conditions improve.

The reproductive characteristics of arid-zone dasyurids are only vaguely known, and, when fully investigated, promise extremely interesting information. The sparse data that are available (summarized in Morton 1982a; Lee, et al.1982; Lee and Cockburn 1985) suggest that the Channel Country species likely are seasonal rather than opportunistic (in response to rainfall) breeders, and that they typically produce two small, consecutive litters during the winter to summer season, regardless of rainfall-induced insect abundance. Thus, if food supply is sufficiently low to result in loss of the first litter there is a possibility that conditions might improve enough for the second to survive. This is a very conservative, if not risky, strategy for life in a region of great unpredictability, yet it apparently pays off, judging from the relatively high species diversity of arid-zone dasyurids. Morton (1982a) suggests that this payoff is related to the very short, ca. 2-year life spans of most arid-zone dasyurids, which preclude their waiting for the "right conditions" to stimulate opportunistic breeding. In the Channel Country of SW Queensland, sufficient rainfall resulting in high rates of vegetative growth (and presumably flushes of insect prey) occurs on the average only once or twice in 10-12 years, far exceeding the 2-year generational life spans of these dasyurids. Thus their short life spans demand that they attempt breeding on a predictable schedule in hopes that at least some of them will produce young coincident with a localized adequate food supply until the next rainfall event occurs, allowing the entire population to rebuild. That winter and spring breeding occurs in the Channel Country is consistent with my observations at Sandringham of females of both the Kultarr (*Antechinomys laniger*) and the Kowari (*Dasyuroides byrnei*) bearing pouch young in August 1966 (midwinter) and October 1974 (early spring), respectively. Both of these periods happened to follow unusually heavy and late monsoonal rains and flooding, with resulting flushes of vegetation and insects.

The final group of sedentary animals that inhabit the Channel Country, whose adaptive responses to this challenging environment I will discuss, is the native placental rodents. We have collected individuals belonging to five species on Sandringham Station, four of which have been studied extensively in the laboratory in studies of energy and water regulation (Table 4). Perhaps representative of the relative abundances of these species on Sandringham, are the numbers taken by one of their chief native predators, the Barn Owl (*Asio alba*), as revealed by analyses of their regurgitated skeletal remains in the form of owl pellets (Table 5, after Morton et al. 1977). These same five rodent species also occur in Barn Owl pellets collected at the ruins of Old Roseberth Station (ibid.), about 200 km south of Sandringham and 35 km north of Birdsville, leading me to believe that these species are widespread throughout the Channel Country. While owl pellets likely reveal what small prey species exist in an area and its several habitats, the cautionary note must be added that owls, like humans, probably have dietary and shopping place favorites, imposing an unknown bias on absolute numbers of each prey species taken. In addition, some prey species are likely more readily caught than are others.

Table 4. Native placental rodent species collected on Sandringham Station, southwestern Queensland, their body masses, and status of physiological investigation (after MacMillen and Lee 1969 & 1970; MacMillen, et al. 1972; Collins and Bradshaw 1973).

Species	Body Mass (g)	Status
Leggadina forresti	16	Unknown
Notomys alexis	29	Energy & Water Metabolism
Notomys cervinus	34	Energy & Water Metabolism
Pseudomys hermannsburgensis	12	Energy & Water Metabolism
Rattus villosissimus	134	Energy & Water Metabolism

Table 5. Native placental rodent species and their abundances in the regurgitated pellets of Barn Owls (*Asio alba*) on Sandringham Station, southwestern Queensland, 1969 and 1970 (after Morton, et al. 1977)

Species	Individuals/156 pellets, 1969	Individuals/137 pellets, 1970
Leggadina forresti	23	13
Notomys alexis	42	83
Notomys cervinus	49	30
Pseudomys hermannsburgensis	12	13
Rattus villosissimus	92	67

To my knowledge, practically nothing is known of the adaptive responses of Forrest's Mouse (*Leggadina forresti*), other than that it is an inhabitant of some of the harshest deserts of inland Australia (Strahan 1995). As it is closely related to, and shares the same habitats with the Spinifex Hopping-mouse (*Notomys alexis*), the Fawn Hopping-mouse (*N. cervinus*), and the Sandy Inland Mouse (*Pseudomys hermannsburgensis*), which I shall discuss below, I will presume that Forrest's Mouse lives a similar life style. The Long-haired Rat (*Rattus villosissimus*) lives such a different life style that I will devote a separate discussion to its existence in a following section on plaguing animals.

Following our initial visit to Sandringham Station in 1966 Tony Lee, Russ Baudinette and I undertook a comprehensive laboratory investigation of water and energy metabolism in the remaining three rodent species identified in Table 4: the Spinifex Hopping-mouse, the Fawn Hopping-mouse, and the Sandy Inland Mouse, all of which occur on and around Sandringham. These studies and the references that pertain to them are discussed in some detail in Chapter 1, so here I shall attempt to incorporate their more salient features into a composite picture of how these small arid-adapted, nocturnally active, and sedentary animals make a living. The Sandy Inland Mouse and Spinifex Hopping-mouse have similar broad distributions throughout arid Australia, extending from SW Queensland to the Western Australian coast, while the Fawn Hopping-mouse has a much more restricted distribution confined to the Lake Eyre Basin (Strahan 1995). This restricted distribution, and its relative abundance (see Tables 1, 2, and 5) suggest that the Fawn Hopping-mouse is unusually suited to the conditions of sustained drought alternating with unpredictably spaced, short periods of torrential rainfall and flooding typical of the region. Where the three species are sympatric in the Channel Country, the Fawn Hopping-mouse concentrates its surface activity on clay pans and gibber, and only on the darkest of nights, while the Spinifex Hopping-mouse and Sandy Inland Mouse seem to concentrate their nocturnal activities in the proximity of sand hills. In these habitats, the Fawn Hopping-mouse, according to my observations, has persisted in seemingly stable, fairly abundant populations over a 26-year period between 1966 and 1992. Over this same period, the Spinifex Hopping-mouse and the Sandy Inland Mouse have occurred only sparsely (Tables 1 & 2). The Barn Owl pellet analyses suggest that these latter two species are more abundant than human observation reveals (Table 5). Our studies have shown that all three rodent species are independent of free water in the moderate laboratory environment, although all will drink water when available or eat succulent plant material, another abundant source of water. When water is withheld, these animals can subsist exclusively on the water of metabolism together with the small amount absorbed in their diet of seeds. Under these conditions of water deprivation, all three species conserve water by producing extremely concentrated urine, with the Spinifex Hopping-mouse and the Sandy Inland Mouse having produced the most concentrated urines yet measured in mammals. They will also freely eat insects and other arthropods, rich in water. As they are nocturnally active, spending their days in underground burrows where temperature and humidity conditions are moderate, they avoid the more demanding daytime surface conditions that would place them in jeopardy. The Fawn Hopping-mouse has the added advantage of being able to process unusually salty water, suiting it even better to its preferred clay pan habitat, where many of the succulent plants it might eat are halophytic (salt-bearing). The infrequent rainwater that collects there dissolves the salts that have accumulated over decades from previous runoff, leaving the likelihood of very salty drinking water, which this species can tolerate. How the Fawn Hopping-mouse copes with complete but temporary inundation of this habitat during the sporadic, massive flooding has yet to be revealed. Clearly though, they are extremely successful, at least on Sandringham Station, for there they represent an abundant, consistently present rodent. Part of their success can probably be attributed to timing their activity periods to coincide with the darkest phase of the moon and period of the night when the ability to evade visual detection by nocturnal predators would be greatest. During new moon on the evenings of 18-20 October 1974, we observed that Fawn Hopping-mice commenced surface activity an hour later each evening, coinciding approximately with moonset. While they may fall prey occasionally to small native carnivorous marsupials (e.g., *Kowari)*, their chief natural predators are likely owls, that sense prey by sound rather than sight, and to which they would be particularly vulnerable in their preferred open habitat, as the data from owl pellets in Table 5 reveal. In early June 1992, while netting Fawn Hopping-mice, we observed many Barn Owls foraging among the clay pans and gibber. Just how long the Fawn Hopping-mouse will remain abundant on Sandringham remains to be seen though, for also in June 1992 we observed for the first time introduced European foxes hunting in hopping-mouse habitat, and joining what appeared to be an increasing number of feral cats. Their combined presence must certainly intensify the predation pressure on these native rodents, as well as on the small marsupials (see Low 2002 for discussion).

There is adequate information available on reproduction in the Spinifex Hopping-mouse and the Sandy Inland Mouse (Breed 1982), and substantially less for the Fawn Hopping-mouse and Forrest's Mouse. All appear to be opportunistic breeders, with pregnant females and subsequent young appearing following sufficient rainfall to stimulate new vegetative growth, apparently independent of season. I noted pregnant and/or lactating female Fawn Hopping-mice in the Channel Country in August 1966 and June 1992, both periods of intense vegetative growth following rains and flooding after drought. Of seven female Fawn Hopping-mice captured at Sandringham in June 1992, four were conspicuously pregnant. It makes considerable sense that reproduction in these four species of arid-adapted mice appears to be triggered by rainfall events, or their resultant renewed vegetative growth, for it is these events that also stimulate renewed production of food energy in the form of seeds, vegetation, and insects that will sustain mouse populations until the next rainfall event. It will not be surprising, and only research will reveal it, to find that their reproduction is triggered by a chemical compound produced by new growth of recently germinated plants, which they consume. Such a plant-produced chemical has been demonstrated to trigger reproduction in field populations of both male and female Montane Voles (*Microtus montana*) in Utah, independent of season (Berger, et al. 1981).

A lifestyle that is somewhat intermediate between sedentary and nomadic that occurs periodically (and unpredictably) in central Australia, and which I have witnessed in the Channel Country, is plaguing. Species that practice this lifestyle are sedentary during periods of drought with small populations isolated in several refugia near permanent waterholes that support the succulent vegetation upon which these animals depend for existence. However, following the infrequent monsoonal summer rains and flooding that occur periodically throughout the Lake Eyre Basin, with ensuing widespread but temporary flushes of succulent annual herbs and grasses, these latent waterhole populations of animals may erupt in a flurry of reproductive activity, yielding an ever-increasing number of progeny and generations that spread in both abundance and range, so long as favorable conditions exist. In a sense, even though individuals may remain somewhat sedentary, the population as a whole becomes nomadic as it increases its abundance and range dramatically, albeit temporarily. As the more common condition of drought begins to develop again, these now widespread populations reduce reproductive output in outlying areas, undergo high rates of mortality due to predation, starvation, and water restriction, and gradually shrink their ever-decreasing populations back to the proximity of permanent waterholes. The three animal species I have witnessed in the Channel Country that practice this lifestyle are the Plague Locust (*Chortoicetes terminifera*), the Plague Rat or Long-haired Rat (*Rattus villosissimus*), and one of the Plague Rat's chief predators, the Letter-winged Kite (*Elanus scriptus*).

The only locust plague I have witnessed was conspicuous, but not very extensive. Terry Dawson, Betty Eidemiller, and I passed through it in October 1974 in the Channel Country between Quilpie on the Bulloo River and Windorah on Cooper Creek in southwestern Queensland, on our way to Sandringham Station. The locusts were extremely abundant for this short phase of our trip, and, while our sighting them followed a period of extensive flooding, we saw few others elsewhere, and I am unaware whether this was the start or end of a plague. Other than this limited observation, I will comment no further about this organism whose life history must be fascinating.

The Plague Rat is much better known, and has become part of Australia's history for it was encountered by members of the Burke and Wills exploratory party while traversing the southern part of the Channel Country in 1860 and 1861. William John Wills, who kept a thorough log of the expedition up to the time of his death on Cooper Creek, first mentions the Plague Rat in his entry of 11 November 1860. He indicates that upon first arriving at Cooper Creek at a place just east of what is now Nappa Merrie Station, Queensland, the party was descended upon by a plague of rats (Moorehead 1985). Wills also notes on the same day that Cooper Creek showed signs of recent flooding (ibid.), likely the stimulus that set off the plague. Wills last mentions the presence of rats, which he was attempting to catch to eat while languishing near today's Innamincka, South Australia, on Cooper's Creek, on 11 May 1861, shortly before his death there in late June (ibid.). Hermann Beckler, a member of a support party that unsuccessfully attempted to rescue Burke and Wills, also comments on the presence of plaguing rats (Beckler 1993). Beckler was a physician and botanical collector, and an astute naturalist. He (ibid.) describes thousands of rats being present between 1 March and 25 April 1861 at several swampy locations, including "Duroado" or "Torowoto" near what is today Tibooburra, New South Wales, "Puria Creek" somewhat to the north, "Kurliatto Creek" on the present New South Wales/Queensland border, and "Bulla", which apparently was the terminus of the Bulloo River in Queensland (see map 3). It is likely that all of Wills' and Beckler's references apply to a single widespread and long-lasting (at least six months) outburst of the Plague Rat.

A much more recent, but apparently even greater plague of the Long-haired Rat is reported by Newsome and Corbett (1975). It followed and/or coincided with unusually widespread and long-lasting summer monsoonal rains in 1966-'67, and again in summer of 1967-'68. A rat plague, or several that coalesced, continued for at least 18 months, forming what appeared to be a single, gigantic crescent-shaped population that extended from north-central Northern Territory southeast into western Queensland, and then south to beyond Birdsville into the Cooper Creek drainage and apparently as far as Nappa Merrie, Queensland and Innamincka, South Australia, where W.J. Wills had reported them over a century earlier. This rat plague of 1966-‘68 was distributed over a linear distance of about 1800 km (1125 mi). The summer of 1968-’69 was accompanied by weak monsoons, followed by continuing drought and poor monsoons in 1969-’70, with the country drying out, and a gradual disappearance of Plague Rats over most of their former range. While we saw no Plague Rats at Sandringham Station in August and October 1966, the Station Managers, Reg and Marg Arthur, reported that the rats had been present in plague proportions a few weeks before our initial arrival, but apparently had withdrawn. However, the rats were conspicuously present and plaguing during our subsequent visits to Sandringham in both 1974 and 1992, periods that followed substantial summer monsoons. On those occasions they were by far the most numerous native mammals present on the station as revealed both by trapping and spotlighting (Tables 1 & 2), and we recorded them at many different station localities, as well as in the vicinity of Glengyle Station, about 100 km to the south. In addition the Plague Rat constitutes the single most abundant mammal present in Barn Owl pellets collected at Sandringham Station in 1969 and 1970 (Table 5), and at the Old Roseberth Station ruins just north of Birdsville in 1966 (Morton et al. 1977).

To my knowledge, only a single study has been devoted to physiological adjustments of the Plague Rat, and this was confined to investigations of metabolic rate, body temperature, and evaporative water loss (Collins and Bradshaw 1973). I am unaware of any studies of tolerances to water restriction and urine concentrating capacities. The one study (ibid.) reveals little out of the ordinary, except the Plague Rat does have a lowered metabolic rate and a reduced body temperature, both of which might serve as energy conserving mechanisms during periods when energy resources are dwindling. The animals apparently are omnivorous, and eat nearly anything organic, typical of opportunistic species. Judging from their population increases when lush vegetation abounds, and their decline when the country becomes desiccated, I would guess that their water conserving mechanisms are modest and that their dependence upon succulent vegetation and/or surface water to meet their water needs is substantial, but only further study will reveal this with certainty.

The data on reproduction in the Plague Rat are rather sparse, too, but it is apparent that they are prolific breeders when monsoonal rains and flooding result in flushes of desert vegetation, providing food, water, and cover for a rapidly expanding population. It is equally apparent that reproduction ceases as the country dries up during an ensuing drought, and only those animals that remain near permanent water and lush vegetation survive and apparently continue reproducing to provide the raw material for the next plague. This species is an ideal candidate to examine the environmental cues that turn on and turn off reproduction, and the role played by plant chemical compounds.

Thus, the Plague Rat has evolved a "Boom or Bust" strategy by which it rapidly saturates vast stretches of central Australia and the Channel Country when that country provides a temporary mesic environment resulting from widespread summer monsoons and flooding. During drought, it retracts in population size and distribution, restricting it to the few remaining mesic refugia that exist. It achieves this strategy by undergoing massive population fluctuations, in which continuing species survival is apparently coincident with death of the majority of individuals, a strange but effective way for the species to make a living!

The Spinifex Hopping-mouse (*Notomys alexis*) is also said to plague after monsoons, although we have seen no signs of this phenomenon in the Channel Country, even though conditions appeared ideal. Newsome and Corbett (1975) report on a plague of Spinifex Hopping-mice near Erldunda, Northern Territory, about 200 km south of Alice Springs, following heavy rains in 1968. In addition, Steve Morton and Pip Masters have commented (personal communication) on more recent plagues of this hopping-mouse near Alice Springs and in the Tanami Desert to the northeast. What stimulates a plague in one place but not in another with similar climatic patterns remains a mystery.

One additional animal that could fall into this category of plaguing species is the Letter-winged Kite (*Elanus scriptus*), a pigeon-sized raptorial bird whose preferred prey is the Plague Rat. The rat likely meets both the energy and water needs of the kite. In this species, breeding is closely associated with exploitation of plagues of the rat (Schodde 1982), with kite numbers and geographic range following closely those of the rat in its population fluctuations. During the rat plague we witnessed on Sandringham Station in June 1992, two pairs of Letter-winged Kites were observed nesting, each in a separate small Coolibah Tree next to the Twelve-mile Waterhole, east of the Station homestead. The ground under the nests was littered with discarded Plague Rat remains, especially parts bearing the tail, anus, and rectum, which the kites understandably might find distasteful! Ian took great delight in referring to them as "*Rattus* rear-ends"! Coinciding with these observations were sightings of numerous runways through the grass-covered, nearby sand hill, attributable to the abundant Plague Rats, which trapping revealed to be present (19 rats in 50 traps in a single night).

The final Life-style category prevalent in the Channel Country to be addressed is that of nomads: animals that roam nearly continually or periodically in search of food and water, commodities whose occurrence and abundance vary in space and time. Among these nomadic species are better than 50 per cent of Australia's arid-zone flying birds (Schodde 1982); the flightless Emu (*Dromaius novaehollandiae*) (ibid.); the larger kangaroos: the Euro (*Macropus robustus*), and the Red Kangaroo (*M. rufus*) (Strahan 1995); and probably the insectivorous bats (Microchiroptera). The Emu and the kangaroos are large fleet-footed survivors of the period of megafaunal extinction, during which their slower-moving, more cumbersome relatives vanished, apparently through direct and/or indirect interactions with early Aboriginals (see Chapter 8; Flannery 1995). Two of these nomads, the Emu and the Red Kangaroo, are so revered by Australians that they were adopted as major features of the national coat of arms in 1908 (Younger 1988).

Very little is known about the adaptations of individual species of Australian bats to aridity (Baverstock 1982). As the arid-zone bats are active at night, seek shelter in moderate microenvironments by day, and eat juicy insects, it is likely that they can meet both energy and water needs as long as their food is available. During drought it is likely that their insect food is present primarily near permanent water, but following flooding and/or high rainfall, insects and bats appear to expand their activities to include some of the rare flushes of vegetation beyond the confines of standing water (Williams and Dickman 2004). Therefore, the bats must follow their food supplies wherever they occur, as insect populations expand and recede. While an individual bat may not be as geographically restricted as small rodents or marsupials with which it co-occurs, its small size and energetically expensive flight nevertheless constrain its mobility much more than that of other nomadic animals, with the possible exception of small birds. Probably, like their northern hemispheric counterparts, Australia's arid-zone bats employ torpor during cool winter days as an energy-conserving measure, and at least some species may hibernate in shelters during the winter (see Geiser 2004), while others may be migratory. Kemper (1990) cites the occurrence of eleven bat species in the northeastern portion of South Australia, near the southern extremity of the Channel Country, and I have seen a few unidentified individuals flying in the evening around Sandringham Station. A recent study conducted on the eastern edge of the Simpson Desert, about 100 km northeast of Sandringham Station, clarifies further bat species presence and habitat use during the high rainfall year of 2000 (Williams and Dickman 2004). Using bat vocalization detectors at three sites, the authors (ibid.) positively identified seven species of bats and tentatively identified three more, for a total of ten. They (ibid.) report maximal vocalization activity during warmer spring and summer periods, and reduced but detectable activity in midwinter. At least part of this midwinter reduction in activity may be due to the utilization of prolonged torpor, as four of the seven bat species positively identified by Williams and Dickman (ibid.) have been demonstrated to employ torpor (Geiser 2004). In addition, Williams and Dickman (ibid.) found that, even during a high rainfall year, bat activity was concentrated near oases, where water, insects, and trees for shelter co-occur; this pattern likely is accentuated during drought for those bats that don't migrate out of the area. Beyond this, little is known and much remains to be learned about these mysterious flying mammals.

Kangaroos are large, conspicuous, fairly abundant and broad ranging marsupials, bipedal when running, and as stereotypically Australian as Foster's Lager beer. Probably because of these attributes, they are the most widely studied of Australia's native mammals. Much of what is known of them is summarized in Terry Dawson's fine, recent book "Kangaroos" (Dawson 1998); Terry has devoted much of his professional career to revealing the intricacies of their physiology, and how these are intertwined with their ecology to make them particularly well suited to central Australia's unpredictable environment. The Euro and the Red Kangaroo are broadly distributed over semi-arid and arid Australia (Strahan 1995). While they overlap in size, the Red Kangaroo is typically the larger of the two species, and males of a species are generally larger than the females (ibid.). The Red Kangaroo has been much more widely studied than the Euro, but where studies have concentrated on both, they appear rather similar in adaptive makeup (Dawson 1998; Denny 1982). Therefore, while many of the comments I make below stem from studies of Red Kangaroos, I will assume they also apply to Euros until proven otherwise. In addition, to my knowledge, the only kangaroos I personally observed in the Channel Country were Reds, and, while conspicuous when present, they were only moderately abundant. Euros prefer rocky and hilly habitats, which occur only sparingly in the Channel Country and not at all on Sandringham Station.

I am classifying kangaroos as nomads that follow the occurrence of their favored foods, green grass and herbs. This nomadism is facultative, depending upon climatic conditions that determine the abundance, distribution, and quality of food and water. Studies indicate that by preference kangaroos inhabit distinct home ranges, discrete areas that provide the resources necessary for daily existence (Dawson 1998; McCullough and McCullough 2000). In addition, there appears to be a negative relationship between home range area and habitat productivity, which in turn is a function of rainfall; that is, as conditions worsen, progressively more area is required to support an individual. If drought persists, kangaroos will abandon established home ranges, and seek more favorable conditions elsewhere, as nomads, until new home ranges can be established that will again meet their needs (ibid.; ibid.). Certainly, during periods of flooding, kangaroos must abandon parts if not all of their home ranges while seeking temporary refuge on higher terrain above flood levels.

Kangaroos are herbivorous animals that prefer grasses and herbaceous plants, and, like their northern hemisphere counterparts, deer and elk, have all of their food-processing machinery geared to this diet. As in deer and elk, the dentition of kangaroos consists of incisors up front that are used for clipping and pulling up preferred plants; behind the incisors is a pronounced gap with no teeth, followed by a row of molariform cheek teeth on each jaw that are used for grinding the plant material into small fragments prior to swallowing. Indeed, a kangaroo skull superficially looks very much like that of a deer or elk. The digestive process also resembles that of their northern hemisphere counterparts in that kangaroos possess a greatly enlarged stomach that serves as a microbial fermentation chamber to render otherwise indigestible plant fibers into digestible energy sources through microbial action. In the stomach, the microbes convert plant fiber and cellular material into more readily digestible foodstuffs that are absorbed as they pass through the kangaroo's digestive tract. In addition, the microbes produce vitamins and convert nitrogen- containing compounds into proteins that the kangaroo utilizes. Finally, during periods of low nitrogen availability kangaroos are capable of redirecting the nitrogenous waste product urea from urine production by the kidney back into the bloodstream; rather than being eliminated in urine, this urea is then secreted into the saliva and swallowed, passing into the stomach where it is converted by microbial action into usable protein for the kangaroo. In combination, the several elements of this food-processing machinery mean that kangaroos, during periods of drought when only low quality grass is available, are unusually efficient in meeting their nutritional needs. A measure of this efficiency is seen in studies by Terry Dawson (1998) in which he compared maintenance food requirements of caged Red Kangaroos and Euros with those of domestic sheep and goats (also microbial foregut fermenters), while subsisting on alfalfa (AKA lucerne) hay. While the fraction of hay digested by the kangaroos was lower than that of the sheep and goats, their mass-relative daily requirement was less than that of the domestic animals. In part, this lower daily maintenance requirement may be attributed to the inherently lower marsupial metabolic rate (see discussion below), but this digestive process is nevertheless very efficient.

As mentioned in the Introduction and in the sentence above, Australian marsupials have resting metabolic rates that are about two-thirds those of placental mammals of the same size, and so are inherently conservative in their energy expenses and requirements while at rest in thermally benign environments (Dawson and Hulbert 1970; MacMillen and Nelson 1969). However kangaroos can be extremely mobile animals, and mobility carries with it additional energy costs that intuition predicts should be positively related to the speed and distance (or duration) of movement, as has been demonstrated in numerous placental mammals across a broad size range (Taylor, et al. 1970). This study of cost of locomotion in placental mammals, however, was conducted on animals running quadrupedally. However, kangaroos running at moderate and fast speeds run bipedally, in a more erect stance with the forelimbs tucked, hopping on the hind legs, with the tail serving as a balancing device; while moving at slow speeds kangaroos are pentapedal, using all four legs along with the tail as a propping device. The question arose whether the cost of locomotion in bipedal kangaroos followed the rules predicted from studies of quadrupedal placental mammals while running. In 1973, Terry Dawson transported several Red Kangaroos to the laboratory of Dick Taylor at Harvard University, and the two commenced a study of their costs of locomotion while walking and running on a treadmill at various speeds. Their results (Dawson and Taylor 1973; and summarized in Dawson 1998) were surprising. While walking pentapedally, the energetic cost of locomotion of the kangaroo exceeds that of similar sized quadrupeds, but when bipedal hopping commences the cost of locomotion remains stable as running speed increases, up to about 30 km/hr. At moderate and high speeds (15-50 km/hr, within the measured bipedal running speeds of Red Kangaroos), the measured or projected cost of locomotion remains less than that of a quadrupedal mammal of the same size. While running bipedally at its most "comfortable" running speed of 20-25 km/hr, a Red Kangaroo requires about 60 percent of the energy it would cost the same-sized quadruped to run at that speed (Dawson ibid.). This energy saving during bipedal hopping, results from storing some of the energy of momentum from each hop temporarily in the elastic elements of the hind limbs, and then releasing that energy to help power the next hop. Biewener and Baudinette (1995) have shown that the smaller Tammar Wallaby (*Macropus eugenii*) while running bipedally stores and recovers up to 60 percent of the energy required to hop in the Achilles tendons of the hind limbs. Such an energy saving while hopping, together with that resulting from the reduced resting metabolic rate, would serve kangaroos particularly well during periods of drought when vegetation becomes progressively scarcer, drier, and of lower quality. This period is when kangaroos are forced to travel greater distances between feeding and drinking sites, or abandon established home ranges altogether while traveling long distances in search of better conditions elsewhere.

With respect to their water needs kangaroos are very economical, although they do require the ingestion of water either in the form of succulent vegetation or standing water when the vegetation becomes drier. Terry Dawson and colleagues (Dawson 1998) have demonstrated with both laboratory and field observations that Red Kangaroos from western New South Wales can maintain water balance without drinking water while eating their preferred food, fresh succulent grass; during the cooler months of the year they rarely drink while this food is available. Under hotter, drier conditions, they (ibid.) found that over a 12-day period only half of 50 radio-collared Red Kangaroos and one-third of 100 collared Euros returned to a water source to drink. Under similar hot and dry conditions at another site in western New South Wales, Dale and Yvette McCullough (McCullough and McCullough 2000) observed that Red Kangaroos restricted their activity, including feeding, almost exclusively to the cooler nocturnal hours, spending the daytime in the shade of shrubs and trees avoiding direct solar radiation, thereby minimizing expenditures of water for evaporative cooling. During colder periods of the year, although still tending toward nocturnality, their activity extended considerably more into the daylight hours. Water balance is further enhanced during hot and dry conditions through the production of quite concentrated urine, thereby reducing water lost by this route (Dawson 1998). This ability to concentrate urine also makes possible the consumption of succulent halophytic plants, employing the kidneys to eliminate excess salts. And the recycling of urea from the kidneys back into the bloodstream and thence to the saliva for microbial action precludes the additional necessity to expend still more water as urine in nitrogenous waste elimination. Thus, it can be seen that kangaroos, while dependent on exogenous water, are extremely efficient in both energy and water regulation.

The final aspect of the kangaroo's adaptive repertoire I will consider, and one that contributes still further to their conspicuous success in an inhospitable and unpredictable environment, is their unusual suite of reproductive attributes. Unless otherwise stated, the information below is gleaned from Dawson (1998). Both Red Kangaroo and Euro males and females breed throughout the year when conditions are good, but cease breeding as conditions worsen during drought. Females of these kangaroos also undergo postpartum estrus and mating, with fertilization producing a new developing embryo shortly after the birth of the previous newborn. However, as long as the previous newborn remains in the maternal pouch, attached to a teat and suckling, hormonal signals induce the second embryo to enter diapause shortly after fertilization, in which further development ceases. This diapause persists until the older sibling develops to the point where it begins to spend time out of the pouch, relying less on maternal milk and more on external herbivory, or it perishes for some reason. In either event, the diapausing embryo recommences development, and proceeds to subsequent birth. During good conditions, there may be a nearly continuous production of young, with a weaned baby at the heels, another suckling in the pouch, and a diapausing embryo in the uterus. However, during drought Alan Newsome (1965,1966) found that reproduction in Red Kangaroos in central Australia became progressively restricted, depending upon the duration and severity of drought. About one-half of the female Reds became temporarily infertile after a 3-5 month drought, but many of them continued to suckle their young. Growth rate of these young was impeded, presumably due to declining quality of maternal milk and external forage, resulting in deaths of one-half of the pouch young after 1 1/2 to 2 months of drought, and all pouch young died after an 8-month drought. Death of pouch young and cessation of lactation, however, triggered reactivation of development of any pre-existing diapausing embryos, to be followed by their births, postpartum mating of the mothers if fertile, and either survival or death of the resulting young, depending upon external conditions. Thus, a succession of newborn kangaroos may be produced, ensuring that when conditions return to normal, new generations will be available to replace individuals that have died or aged. Like all marsupials, their young are born in an extremely immature, almost embryonic state, following a relatively brief gestation period of about a month. As such, the single newborn young is tiny, constituting only a minute fraction of the mother's body mass, and therefore representing minimal investment by the mother of time, energy, and nutrients, if lost. The primary investment of these commodities comes after birth, in providing the young with milk, care, and transport in the maternal pouch. The magnitude of this investment is directly related to the duration of maternal care, which, in turn, is determined by environmental conditions: when conditions are bad, the young perish, but when good they survive to independence which is achieved in about a year.

Thus it can be seen that kangaroos serve as excellent models of what it takes to succeed, and at times flourish, midst the unpredictable vagaries of the central Australian landscape. As seen above, kangaroos have a combination of: nomadism/sedentariness; use of microbial fermentation in digesting otherwise unpalatable dry grass; low resting and running energetic costs; great economy in water expenditures and needs; and low cost, usually dependable reproduction. The sum of all of these adaptive traits of kangaroos means that they are ideally suited for the unpredictability of climate and ensuing biotic events that so characterize the Channel Country of central Australia, as well as the greater arid zone that surrounds it.

The Emu, another nomad, occurs practically everywhere in Australia, except in densely settled areas, and in Tasmania, where it was exterminated by 1870 (Slater, et al. 2001). It is an omnivorous inhabitant of grasslands, open plains, and savannahs, where it roams broadly in search of a wide variety of food. While the Emu's natural history and behavior are fairly well known, its intractable nature in captivity has limited the number of physiological investigations to a handful. None of these studies has taken place in the Channel Country, where the Emu is only sparsely distributed, but the composite picture gleaned from a variety of studies likely applies to Emus there as well as elsewhere.

Davies (1976; Davies, et al. 1971) has studied the nomadic nature of Emus in Western Australia, where he has found that they migrate seasonally with the occurrence of food provided by winter Mediterranean rains in the south and summer monsoonal rains in the north; one individual marked for identification traveled a distance of 990 km. In another study in semi-arid country of NW New South Wales, Terry Dawson and colleagues (Dawson, et al. 1984) noted that Emus spend most of the daylight hours walking and feeding, but always within daily reach of surface water. During mild winter weather, they tended to drink once a day, and traveled about 10-12 km/day; during hot summer days, at least some birds drank twice a day, and all birds were estimated to travel between 18 and 25 km/day. This daily drinking contrasts with their nomadic neighbors the Red Kangaroo and Euro, that during the same periods rarely returned to water in less than 3-4 days or longer (ibid.), attesting to the much greater reliance of Emus upon frequent access to surface water. I infer from these studies that Emus roam throughout the Channel Country in search of food, but are seldom more than a short walk to water. During summer monsoonal conditions Emus are likely dispersed, as are their watering and feeding choices, but during drought they probably concentrate in the vicinity of the few permanent waterholes that persist, or leave the country altogether.

While Emus apparently prefer to remain near drinking water, they will tolerate, under experimental conditions, several to many days without drinking water while on a dry diet. Over a nine-day period of water restriction, experimental Emus reduced their water loss rates to one-third that of birds provided with water, although they lost a considerable amount of body mass and were in a state of negative water balance during that period (Dawson, et al. 1983). This tolerance of periods of water deprivation may serve them well during drought while seeking more favorable conditions elsewhere. One of the more remarkable features with regard to Emus' tolerances of water deprivation was demonstrated in an incubating adult. While the female lays the eggs, the male Emu assumes all incubation and care-of-fledgling duties. The incubation period is about 8 weeks, during which the attending male seldom eats or drinks, tolerating a lengthy famine and sustaining substantial weight loss. Terry Dawson (Dawson, et al. 1984) observed that an incubating male reduced its rate of water loss to one-fifth that of the same bird when not incubating, but his observations did not reveal the mechanism by which this was accomplished.

Birds in general, including the Emu, void much of their nitrogenous waste as osmotically inert uric acid and urate salts, with minimal loss of water (Gordon, et al. 1977). However, other metabolic wastes must be eliminated in solution, taking with them water in the form of urine. A measure of water economy in birds is the extent to which they can concentrate urine, thereby conserving water internally that otherwise would be expended (ibid.). Skadhauge (1974) has measured the urine concentrating capacity of a variety of Australian birds while undergoing drinking-water deprivation, including the Emu and several other inhabitants of the arid zone. In general, the arid-zone species (e.g., Zebra Finch, Bourke's Parrot, and Galah) were good urine concentrators, but the Emu produced the least concentrated urine of the ten species studied (ibid.), and about equivalent to that of the domestic chicken and turkey when water deprived (Willoughby and Peaker 1979). Thus, without further information, I must conclude that the Emu has few special physiological tricks to equip it for life in the Channel Country, and that its large size, omnivorous diet, and behavioral predilection to roaming imbue it with sufficiently high mobility to nearly always keep it in the proximity of food and water.

The final group of nomadic animals to be mentioned is also the most numerous and diverse among the vertebrates, yet still not fully studied and understood. These are the flying birds, that carry with them the added burden of being largely day-active, and spend the inactive periods (at rest or at night) on the desert surface, facing directly the demands of the desert environment. They do not have the luxury of escaping to more moderate underground burrows, as do many of their reptilian and mammalian neighbors. Well over 50 percent of the flying bird species that occur in Australia's arid zone are nomadic (Schodde 1982). Of these, many are water and shore birds whose movements always keep them close to water, no matter how scarce it may be; but an even larger number are land birds whose ties to surface water are less conspicuous (ibid.). All of these nomads, however, are strong fliers, capable of moving long distances to find the ever-decreasing aquatic habitats during drought, or of moving progressively greater distances between feeding and watering places that usually become more separated during drought. The aquatic habitats of permanent and even ephemeral waterholes apparently supply the water and shore birds with all the necessities of existence, and, for them, where there is water also there are their aquatic foods. When this preferred habitat deteriorates, they have the luxury of flying off to the next waterhole that will meet their needs, even if it is a great distance away. Land birds, however, frequently find their preferred foods at varying distances from drinking water. For insectivorous, carnivorous, and nectarivorous (nectar-feeding) birds, whose diets contain appreciable water, this may not pose a difficulty even during drought, provided they find ample food. However, for birds that rely on drier foods, especially seeds, a potential problem exists, particularly during drought when rich supplies of seeds may be far removed from sources of drinking water. Concentrations of seeds, most frequently grass seeds, may be scattered patchily across broad expanses in the Australian arid zone (Morton and Davies 1983), due to the patchy occurrence of rainfall even during prolonged drought. Localized, unpredictable rainfall, sufficient to induce seed germination, vegetative growth, maturation and seed-set, but not lasting surface water for drinking, is common in central Australia (personal observation). Among birds, the arid-zone granivorous species, which include doves and pigeons, several finches, and a variety of parrots, meet head on the most intense challenges of Australia's desert environment. Some of these have been studied fairly thoroughly, while others remain relatively unknown.

The late Alden Miller, an eminent University of California (Berkeley) ornithologist, first noted in print that birds in general have several important, inherent avian traits that pre-adapt them for successfully occupying deserts, at least temporarily (Miller 1963). These traits include: 1) high body temperatures and the ability to tolerate mild hyperthermia, such that body temperatures usually exceed environmental temperatures even on hot days, precluding the necessity of costly evaporative cooling for thermoregulation; 2) nitrogenous wastes that are excreted in the form of the insoluble precipitate uric acid, with minimal expenditure of water; and 3) high mobility via flight, enabling birds to travel long distances to find water, or to escape intolerable conditions. Add to these avian qualities a few additional specialized adjustments, and it becomes not only theoretically possible, but actually probable, for at least certain birds to become desert specialists, occupying permanently some of the harshest deserts of the world. Paradoxically, however, among these birds are granivorous ones, eating at times some of the driest food available. Furthermore, among these granivores are birds that appear to be completely independent of drinking water, at least under moderate laboratory conditions. In 1972, my mentor George Bartholomew reviewed what was known at that time about the physiological attributes of these water-independent seed eating birds (Bartholomew 1972). He postulated that small size, and a favorable relationship between evaporative water loss (their major avenue of water loss) and metabolic water production (the major avenue of water input), particularly at lower environmental temperatures, promoted positive water balance in the absence of drinking water. Among these small, water-independent birds were two that inhabit the Australian arid zone, the Zebra Finch (*Poephila guttata*) and the Budgerigar (*Melopsittacus undulatus*); curiously enough, their water independence was first demonstrated in many-generations raised aviary birds in the United States. Shortly thereafter, Lindgren (1973) demonstrated drinking-water independence in another small Australian arid-zone bird, Bourke's Parrot (*Neophema bourkii*), in a study conducted in Western Australia. These studies, additional ones that followed, and my own work on water-independent granivorous rodents (both American and Australian) prompted me to turn my attention to granivorous birds. This attention resulted in the formulation of predictive model of water independence, or lack thereof, in arid-zone granivorous birds (MacMillen 1990).

In 1989 I had the opportunity to test this model on Australian parrots in a collaborative study with Russ Baudinette at Flinders University of South Australia (MacMillen and Baudinette 1993), and both the model and the parrot study are outlined in detail in Chapter 5, above. Briefly, the model states that the main determinant of water regulatory efficiency in granivorous birds is the ratio between evaporative water loss (the major avenue of water output) and metabolic water production (the major avenue of water input). When evaporative water loss is less than metabolic water production, independence of drinking water may be achieved, with the bird operating at maximal water regulatory efficiency. Correlates of these determinants as additional factors influencing water regulatory efficiency include environmental temperature, body size, time of day, and seed composition (i.e., amount of carbohydrate, protein, and fat). The model predicted greatest water regulatory efficiencies at lower environmental temperatures, and a size-related progression, with large birds being the least efficient, regardless of diet, and small birds being the most efficient, particularly if they specialized in eating carbohydrate-rich seeds with high metabolic water yields. We tested the model with six species of Australian arid-zone parrots, all of which I have seen in, or are known to inhabit (Blakers, et al. 1985), the Channel Country; these species ranged in size from the largest, Little Corella (*Cacatua sanguinea*, 391 g), to the smallest, Budgerigar (*Melopsittacus undulatus*, 27 g), with approximately equal size gradations between.

Russ and I found (MacMillen and Baudinette 1993), as the model predicted (and as previous studies had shown), that the small Budgerigar and next smallest Bourke's Parrot (39 g) were indeed drinking-water independent under moderate laboratory conditions while eating millet seeds (approximating native grass seeds in high carbohydrate composition); the larger species all required drinking water to maintain body weight. We found further that water regulatory efficiency in all species was greatest at night, following a period of progressive drinking water restriction, but that it was only the two smallest species that were water independent at moderate and low temperatures. When the laboratory performances of these six parrot species were applied to conditions known to exist in central Australia, including the Channel Country, we concluded the following: 1) as long as carbohydrate-rich seeds were available to eat and assimilate during the colder winter months, the Budgerigar and Bourke's Parrot may remain in positive water balance all night long, and for most of the day, without drinking; during the warmer parts of winter days, a short period of negative water balance might accrue without drinking water, but this water deficit could be repaid with the water surplus that would accumulate overnight, potentially resulting in complete drinking water independence during the winter; 2) during the warmer seasons, and particularly during summer, even these smallest parrots require drinking water to maintain water balance and body weight; 3) parrots larger than about 45 g require drinking water regularly and regardless of season in order to maintain positive water balance, even when operating at peak water regulatory efficiencies. Bird flight energetics are such that larger birds get better fuel mileage and can carry more fuel than smaller ones, and therefore larger birds have greater flight ranges than smaller ones. Thus, during the cool, dry winter season when seed sources are likely to be concentrated in patches that are far removed from each other, and from surface water, the larger parrots have the capacities to fly daily between feeding and watering sources. The smallest parrots with their more limited mobility, on the other hand, possess the potential of remaining in the vicinity of a rich seed patch continually until it is exhausted, and while maintaining both energy and water balance. During the warmer seasons, however, all parrots, regardless of size, become obligate drinkers, which poses no difficulty if summer monsoons occur, as water then is widespread and abundant as are seed sources. It is the long periods of summer drought that are the most demanding, especially for the less mobile smaller parrots who must remain within flight distance of drinking water and a progressively diminishing food supply; the more mobile larger parrots are at an advantage then, and can seek resources at greater distances. All of these parrots may be constrained still further, as their nest sites are in the cavities of trees that occur predominantly along watercourses. During drought, this tree-nesting habit may place the nesting birds near water and, at the same time, perhaps at even a greater distance to a reliable food source. In spite of these demands, parrots flourish in the Australian arid zone, and often are among the most conspicuous and abundant elements of the avifauna there (see photo of Galahs, fig. 17).

Figure 17. A pair of Galahs perched on a dead gum tree snag above a billabong on Mabel Creek Station, near Coober Pedy, South Australia, 1989.

Perhaps more abundant than most parrot species in arid Australia and the Channel Country is the Zebra Finch (*Poephila guttata*), another granivore that occurs nearly everywhere in Australia, except Tasmania (Blakers, et al.1985). This small finch (ca. 12 g) is only about half the size of the smallest Australian parrot (the Budgerigar), and the data available (Cade, et al.1965; MacMillen 1990) indicate that the Zebra Finch is even more efficient in water regulation than is the Budgerigar, as might be anticipated for a still smaller granivorous bird. A question, as yet unanswered, remains in my mind about water regulatory efficiency in granivorous birds: if smaller is better, why are there not other Australian arid-zone parrots that are even smaller than the Budgerigar?

An ambitious study of the drinking habits of Australian arid-zone birds was conducted between 1966 and 1968, by Charles Fisher, Eric Lindgren, and Bill Dawson, working out of the University of Western Australia (Fisher, et al. 1972). The study consisted of dawn to dusk observations during the warmer, drier parts of the year at eleven waterholes scattered throughout the arid zone, four of which were located within the Lake Eyre Basin. All six of the parrot species employed in Russ Baudinette's and my study were observed in theirs, and the Zebra Finch was prevalent in their study, too. They (ibid.) concluded that granivorous birds are the most abundant avian group in the Australian desert, and I contend, therefore, that they are also the most successful. Fisher, et al. (ibid.) state, as an example of this abundance, that during September 1967, the greatest single-day count by species of granivorous birds visiting a waterhole located 44 miles north of Alice Springs, Northern Territory (at the northwestern extremity of the Lake Eyre Basin) was: 4,500 Cockatiels (*Nymphicus hollandicus*, another parrot), 47,000 Budgerigars, 800 Bourke's Parrots, and 17,750 Zebra Finches. The latter three species were all recorded on the same day (21 September 1967), while the Cockatiels were recorded two weeks earlier, with both periods having mild days (ca. 23°C) and cool nights (ca. 3.5°C). They (ibid.) also indicate that this was a very dry period, preceded by seven months of very little rain, and that seeds were abundant in the surrounding area. Even though these observations were preceded by several months of dry conditions, it is probable that this great abundance of granivorous birds resulted from very successful reproduction following abnormally high monsoonal rainfall there in March 1967, as reported in Newsome and Corbett (1975), and described at the beginning of this Chapter. These authors (ibid.) indicate that this rainfall was followed for many months by lush and seeding herbage, likely resulting in the abundance of seeds in proximity to the waterhole observed by Fisher, et al. (1972). Given the abundance of birds and seeds, and followed by several months of drying conditions, it is not surprising that great congregations of birds would gather around an isolated waterhole to drink and feed. Thus, it is apparent that if water is available in the proximity of a feeding area, these birds will drink. However, it is our contention (MacMillen and Baudinette 1993) that, when conditions are cool to cold and feeding grounds are far removed from water, at least Budgerigars, Bourke's Parrots, and Zebra Finches do not need to fly to water each day in order to remain in positive water balance as long as they can find seeds. Supporting our position is the statement of Fisher, et al. (1972) that at least Budgerigars and Bourke's Parrots may satisfy their water needs with "possibly a single visit only once every 2 or 3 days". It is somewhat paradoxical that the Zebra Finch, with its even more efficient water regulatory machinery, was observed only in the vicinity of drinking water, which they visit throughout the day, with individuals drinking at least several times a day, regardless of temperature conditions (ibid.).

Fisher, et al.'s (ibid.) study of the drinking habits of Australian arid-zone birds also revealed that several species of doves and pigeons were common visitors to water sources during the day, being outnumbered only by parrots and Zebra Finches. These six species of doves and pigeons also are granivorous, and were observed to drink frequently and avidly (ibid.). All six species occur in the Channel Country (Blakers, et al. 1985), where I have observed most of them. With the exception of the Spinifex Pigeon (*Petrophassa plumifera*), these birds are all strong fliers, and, like the arid-zone parrots, they vary considerably in body size. The Spinifex Pigeon, although of intermediate size, is the most likely candidate among them to show unusual water regulatory capabilities, for it is quail-like in habit, preferring to run along the surface rather than using its more limited flight capacities for other than sudden escape. Studies of thermoregulation in the Spinifex Pigeon revealed nothing exceptional (Dawson and Bennett 1973), and my hope to study its water economy in 1966-67 had to be aborted due to an inability at the time to collect the birds (see Chapter 1). A timely study could/should be conducted across the size range of Australian arid-zone doves and pigeons, including the Spinifex Pigeon, to assess their water regulatory efficiencies, and the extents to which they are dependent upon surface water.

Schodde (1982) has reviewed the reproductive characteristics of the land birds that inhabit Australia's arid zone, and noted that earlier ornithologists (e.g., Serventy and Marshall 1957) felt that breeding was opportunistic, induced by new growth of vegetation and moderate to high temperatures following rain, rather than strict seasonal influences. Indeed, Schodde (ibid.) points out that the arid-zone nomadic birds in particular tend to nest irregularly. However, according to Schodde (ibid.) there is increasing evidence that the short day lengths of winter inhibit reproduction, while the onset of longer spring days brings on gonadal development and the impulse to breed. Breeding will ensue, generally in late spring, sometimes extending into summer and early fall, depending upon other physical and biotic conditions. In dry years, breeding is short, with small clutches, but in wet years it is extended with larger and repeated clutches. These differences result in the irregularity noted above, and depend upon the availability of food, which, in turn is a function of rainfall and temperature. Schodde (ibid.) believes though, that the ultimate trigger governing breeding in these birds is seasonal day length. However, it is the timing and the amount of rain, and the resulting primary and secondary productivity that determine the duration of breeding, and its success.

I truly believe that the Channel Country of the Lake Eyre Basin and the Australian Arid Zone harbors a fauna that is unique among the world's deserts, one that has responded to an equally unique suite of environmental circumstances that includes extremes of prolonged drought truncated unpredictably by summer monsoonal rains and flooding. Over evolutionary time, this has resulted in a remarkably diverse collection of aquatic and terrestrial animals that possess equally diverse strategies for successfully inhabiting this country. I have had the rare privilege of observing many of these animals in action, and, in the essay above, have attempted to translate these observations into descriptions and speculations of how at least some of them make a living. Admittedly, these observations pose as many questions as they do answers, but an important role of scientific investigation is to provide fertile fields for further inquiry by curious minds. Happy future meanderings, the great Australian bush is out there waiting for you!

SUGGESTED FURTHER READING

Barker, W.R., and P.J.M. Greenslade (editors). 1982. Evolution of the Flora and Fauna of Arid Australia. Frewville; Peacock Publications.

Cogger, H.C., and E.E. Cameron (editors). 1984. Arid Australia. Sydney: Australian Museum.

Dawson, T.J. 1998. Kangaroos: Biology of the Largest Marsupials. Ithaca: Cornell University Press.

Dawson, T.J. and M.D.B. Eldridge (editors). 2004. Animal Function in Arid Environments. Australian Mammalogy 26: 109-226.

Chapter 8

WILDLIFE OF THE CENTRE: THE HUMAN DIMENSION

> *A few years since this country abounded with wild animals; now the emu is banished to a long distance and the kangaroo is becoming scarce; to both, the English Greyhound is utterly destructive; it may be long before these animals are altogether exterminated, but their doom is fixed.*
>
> Charles Darwin, January 18, 1836.
> From his Diary of the Voyage of the Beagle
> (Barlow 1933)

The initial human intrusion onto the Australian continent was by Aboriginal people. Their arrival likely was from southeast Asia by way of the Malay Archipelago, then to New Guinea and finally to Australia, with passage made easier by late Pleistocene glacial sea level lowering and numerous land connections that no longer exist. Likely, there was more than one invasive wave of Aboriginals, with the first occurring well before 40,000 years ago (see for example Dayton, 2003). The earliest Aboriginal remains thus far discovered and dated are those of two ritual burials found in dune deposits along the shore of now dry Lake Mungo, located in the semi-arid southwestern portion of New South Wales. These remains were discovered between 1969 and 1974, and were thought at the time to have been at least 30,000 years old (Bowler, et al. 1970, 1976). Recently Alan Thorne and colleagues at Australian National University applied three independent, sophisticated dating techniques (thermoluminescence and optically stimulated luminescence, electron spin resonance, and uranium decay) to one of the remains to obtain a more accurate date; all three techniques revealed nearly identical ages of the remains at 62,000 years ago (Thorne, et al. 1999). This date of 62,000 years lends further credibility to the dates of 50,000 to 60,000 years, as determined with thermoluminescence dating of an occupation site in Arnhem Land, Northern Territory, by Roberts

(Roberts, et al. 1990). However, more recently Bowler and a team of experts representing four other independent laboratories employed a refined optically stimulated luminescence technique to date the sediments in which the two ritual burials were contained (Bowler, et al. 2003). They (ibid.) arrived at nearly identical burial dates of 40.000 years ago for both remains, and an earlier date of 46.000- 50,000 years ago for human-crafted stone implements near the burial sites. These conflicting Lake Mungo dates of 62,000 and 40-50,000 years ago are yet to be finally resolved, but current opinion (Forster and Matsumura 2005) agrees that southern Australia was definitely inhabited by early Aboriginals at least 46,000 years ago. If ancient Aboriginals invaded the Australian continent from the north, as is likely, and had extended their range thousands of kilometers to the southern part of the continent (e.g., Lake Mungo), their original arrival in the north must have been still earlier.

Another line of evidence suggesting a relative time of arrival of Aboriginals onto the Australian continent is seen in cores of sedimentary deposits that formed in lakes over and beyond the last 100,000 years. Such cores have been examined in deposits in the extreme northeast of the continent (Cape York Peninsula, Queensland), and in the extreme southeast (near Canberra, ACT). In addition to mineral sediments, these cores also bear charcoal fragments from nearby wildfires, and pollen grains from the then local flowering plants. These in turn may be used to denote fire frequency and intensity, and floral composition over time. From both places, and seemingly nearly simultaneously, there are sudden marked, continuing increases in the amount of charcoal, and equally sudden changes in floral composition. Both of these sudden events are attributed to the initial arrival of Aboriginals, and their use of bush fires to enhance productivity, and success in foraging and hunting, and the resultant changes in vegetation. Such use of fire has been documented as a common Aboriginal practice (Tindale and Lindsay 1954). The timing of these simultaneous events has been placed somewhere between 38,000 and 125,000 years ago; the former date was assigned by radiocarbon dating methods, and the latter upper limit was assigned by stratigraphic dating (see Mulvaney and Kamminga 1999, for discussion). The 38,000 years date approaches the upper temporal limit of the radiocarbon technique, in a region of increasing inaccuracy; the dates beyond were assigned by stratigraphic extrapolation and interpolation, even more approximate methods. The true dates of the sudden appearances of massive charcoal deposits resulting from the arrivals of Aboriginals, and pollen indicating floral compositional changes, are likely intermediate to the two extremes noted above, but they will remain unknown until more precise methods, such as thermoluminescence, are applied. Once applied, it will not be surprising if the dates correspond well with those recently ascribed to the ancient Aboriginal remains of Lake Mungo.

When Aboriginals began their habitation of arid central Australia is uncertain, but it was at least 22,000 years ago, as radiocarbon dating of materials at the Puritjarra rock shelter, west of Alice Springs, has shown (Smith, 1987). There is indirect evidence that habitation of the arid zone was much earlier, perhaps at least 50,000 years ago (see discussion on extinction of *Genyornis* below). Since the source of the original Aboriginals was most likely from the moist tropics of Malaysia it makes sense that these immigrants would initially seek similar moist regions as they began to occupy Australia, avoiding the more unfamiliar and hostile arid interior. This scenario would place the early Aboriginal dispersals along the coastal fringe, especially along the east coast where mesic conditions exist without interruption along its entirety, including the late Pleistocene land bridge then uniting southeastern Australia with Tasmania. There is indirect evidence that the Lake Eyre Basin, now arid throughout most years, had very predictable summer monsoons in the period from 45,000 to 65,000 years ago, more so than today (Johnson, et al. 1999). This period probably also encompasses the arrival and exploratory dispersals of the earliest Australians, and would favor opportunities during the summer monsoons for at least temporary forays into the interior from the less hostile coastal fringe. One can imagine that, with increasing population size, some of the more intrepid explorers would figure out ways to make a permanent living in the interior, even during the dry winters, and as the summer monsoons became less predictable. The combination of knowing the location of waterholes and seeps even during drought, together with regional nomadism, and the use of wildfire would make an otherwise hostile desert landscape habitable.

These desert people made extensive use of fire, as had their progenitors in more mesic regions and their descendants up to present times, in order to manipulate the landscape consistent with enhancing food procurement (Jones 1969). Certainly, this systematic use of fire, applied in small patches, altered the landscape, but it is believed that the resulting enhanced primary production may have been as beneficial to wildlife as the fires were detrimental, even when allowing for increased hunting success. Such land management practices are consistent with the Aboriginal cultural belief in the Dreaming, in which nothing is created by human beings, it was all there from the start; the Dreaming left behind at various places the creative potency of all the natural species and of human beings (Myers 1991). The Dreaming is an integral part of Aboriginal oral history, passed on from generation to generation. According to mythology, it goes back in time as far as the Aboriginals themselves, but the concept is just as likely to have evolved in Australia while these new pioneering inhabitants adapted to their strange new surroundings. Thus the Dreaming, as interpreted to 20th and 21st century anthropologists, represents at least 50,000 years of conforming progressively to Australia's landscapes. In its presently understood form...."The Dreaming ... denies creative significance to history and human action, just as it denies the erosion of time. It represents all that exists as deriving from a single, unchanging, timeless source. All things have always been the same, forever deriving from the same basic pattern. The Dreaming, which cannot be altered by human action, is the very image of self-direction and the source of a given autonomy in human life" (ibid.). Thus, the entire Aboriginal philosophy is that they, as individuals, are set to reside on the land temporarily, as stewards of the landscape, to ensure that no real changes take place, a philosophy of sustained yield, ensuring their survival. By the use of fires, and a nomadic existence, a sufficiency of resources could nearly always be found somewhere, without long-term landscape degradation. However, there is reason to believe that this philosophy may have been adopted as a result of the initial 1,000 years of occupancy, after a series of trial and error mistakes and successes.

Extinctions of plants and animals are incompatible with the traditional Aboriginal concept of the Dreaming. Yet, Australia, like other major mid-latitude continents, had a late Pleistocene terrestrial vertebrate megafauna that became extinct. Australia's megafauna was made up of giant marsupials, flightless birds, and reptiles. In all, every terrestrial vertebrate species larger than humans was lost, representing at least 60 species of which at least 20 species were kangaroos and their relatives (Flannery 1995, 1999). One of these megafaunal elements was *Genyornis,* a clumsy, flightless bird about twice the size of the modern, swift-footed Emu. Using sophisticated dating techniques (amino acid racemization, accelerator mass spectrometry, radiocarbon, and thermal ionization mass spectrometry) on their eggshells, Miller, et al. (1999) have shown that in three separate regions of south-central Australia (including the arid Lake Eyre Basin) *Genyornis* and the Emu coexisted between 100,000 and 50,000 years ago; at the latter time *Genyornis* became extinct, leaving only the Emu to survive to the present. The timing of this extinction was not long after the widespread appearance of Aboriginals across Australia, and their apparent use of wildfire to manipulate and manage the landscape. Since 50,000 years ago was a period of only modest climatic change, these authors (ibid.) attribute the extinction of *Genyornis* directly and/or indirectly to activities of Aboriginals, through predation and/or fire-induced habitat alteration. By inference, Aboriginals responsible for *Genyornis* extinctions in the Lake Eyre Basin must have inhabited these arid regions by 50,000 years ago.

More recently, an interdisciplinary group of mostly Australian scientists (Roberts, et al. 2001) has examined 28 fossil megafaunal sites that include most major biogeographic and climatic regions of Australia, and all megafaunal taxa (birds, mammals, reptiles). By using a combination of optical luminescence and thorium/uranium dating methods applied to sediment grains immediately associated with articulated fossil skeletons, the team concludes that megafaunal extinctions occurred nearly synchronously across the entire continent and across all climatic regions about 46,400 years ago (95% confidence interval, 51,200 to 39,800 years ago). Further, this period precedes by at least 20,000 years any major climatic event (last glacial maximum) that might have contributed to massive extinctions. Thus these scientists conclude that aboriginal peoples were responsible, either directly through hunting or indirectly thorough habitat alteration (or both), for the disappearance of the Australian megafauna, and within only a few millennia following their arrival in Australia. However, it remains enigmatic that no direct evidence has yet been found that links early Aboriginals as predators with the extinct megafauna they are presumed to have preyed upon (Flannery 1995). Most recently an international team of experts (Cardillo, et al. 2005) has documented that, at least in mammals (and by inference animals in general), a size-threshold of 3 kilograms exists above which there is an intrinsic susceptibility to extinction resulting from human-induced environmental changes. Thus, this study implies further that direct predation by Aboriginals may not have been as important as habitat alteration in contributing to these megafaunal extinctions. These Australian extinction events precede by tens of thousands of years similar megafaunal extinctions in North and South America, Madagascar, and New Zealand, each of which was preceded closely by the arrival of humans, further implicating human actions as the causal factors of megafaunal extinctions rather than any global climatic event.

Following the megafaunal extinctions the central Australian Aboriginals, through their system of habitat enhancement by patch-burning and nomadism, probably should be viewed as trial-and-error wildlife managers as their concept of The Dreaming suggests, rather than as decimators of wildlife. Their form of hunting and gathering began to approach sustained-yield agriculture, a practice yet to be perfected and adopted by most modern European cultures. It is likely, however, that the Aboriginal practice of spot-burning resulted in as yet unknown distributional shifts of organisms and in changes in species densities and composition that differed from the non-burned, pre-Aboriginal setting, notwithstanding the megafaunal extinctions.

Another major impact on the Australian biota induced by pre-European people was the relatively recent introduction of the Dingo or wild dog

(*Canis familiaris dingo*), which quickly became naturalized as the original terrestrial eutherian mammalian predator. The Dingo is not only a vigorous predator but it also may be regarded as a significant competitor of other Australian carnivorous mammals, all of which are (or were) marsupials. The oldest reliably dated Dingo fossils are from rather widespread sites in New South Wales, South Australia, and Western Australia, dated between 3,170 and 3,450 years ago (Corbett 1985; Flood 1989); this broad distribution suggests a probable earlier initial appearance. It is now pretty well agreed (Corbett 1995; Dayton 2003) that the first Dingoes were brought to Australia as pets from the islands of southeast Asia by Asian seafarers about 5,000 years ago; the genetic homogeneity of pure-bred modern Dingoes suggests that they owe their Australian origins to perhaps only a single introduction (Dayton 2003). The absence of the Dingo from Tasmania is attributed to the separation of that island from the mainland through late postglacial Pleistocene increases in sea level prior to the arrival of the Dingo, between about 12,000 and 8,000 years ago (ibid.). Aboriginal rock paintings and fossils tell us that the two largest marsupial predators, the Thylacine (*Thylacinus cynocephalus*) and the Tasmanian Devil (*Sarcophilus harrisii*) were once widespread on the Australian mainland. They survived aboriginal occupation on the mainland, but became extinct prior to the occupation by Europeans in the 1700s; the latest Thylacine fossil from the mainland has been dated at not less than 2,000 years ago, and a mainland Tasmanian Devil sub-fossil at 600 years ago (ibid.). Most authorities agree that their decline and disappearance was due to competitive displacement by the Dingo. Supporting this view was the common presence of both of these marsupial predators in Tasmania, in the absence of Dingoes, when Europeans first arrived. While the Thylacine is now considered extinct in Tasmania, the Devil still survives. Thus, pre-Europeans are implicated only indirectly in the extinctions of these two marsupials on the Australian mainland.

Following the megafaunal extinctions, and after about 50 millennia of relatively benign occupation of Australia by Aboriginals, it has taken hardly more than 200 years of occupation by Europeans to initiate and continue a malignant episode of habitat degradation and species destruction, resulting in numerous extinctions and threatening ever more. Particularly hard hit among animals have been the mammals, especially the small and middle-sized ones, with both marsupials and placental rodents falling victim. Of the vertebrates 9 percent of the known mammal species have suffered extinctions during European occupation, while only 0.4 per cent, 0.1 percent and 0.0 percent of birds, reptiles and amphibians, respectively, have become extinct (Kennedy 1990). The arid zone has been particularly vulnerable, having lost 33 percent of its known mammalian species; arid zone scientists estimate further that 90 percent of all medium-sized desert mammals are either extinct or threatened with extinction (ibid.). Most authorities seem to agree that these massive extinctions, particularly in the fragile Centre, stem largely from habitat degradation imposed by a combination of European grazing and a disruption of Aboriginal land-management practices that relied upon patch-burning. During the process of development of the Australian grazing industry in the Centre most of those Aboriginals that survived the onslaught of pastoral domination were removed from their traditional lands to government and religious settlements, effectively ending thousands of years of accumulated land management practices. Displacing them were Europeans with their myriad introduced herbivorous mammals, including cattle, sheep, horses, donkeys, Camels, goats, pigs, rabbits, rats and mice, many of which have established large feral populations in the Centre (McNight 1976; Low 2002; and see fig.18). Also introduced and now established as feral predators are the domestic cat and the European fox. This horde of large grazers, smaller competitors, and effective predators rapidly resulted in the combination of habitat destruction, declining populations and predation that renders species extinct.

Figure 18. A typical sign along a highway in central Australia warning drivers to be alert for native kangaroos and emus, and the introduced but now established Camels.

The first 100 years of Australian occupation by Europeans likely had little effect on the biota of the Centre, as European settlement was confined initially to equable coastal regions. It was not until the middle of the 19th century that the curious and land-hungry settlers began to penetrate beyond the coastal fringe and into the interior, stimulated at least in part by the discovery of mineral wealth in northern Victoria and western New South Wales. At about the same time intrepid explorers began to penetrate further and further into the Centre, some even managing to traverse it. Among the most adventurous of the lot were the early pastoralists who sought out lands in central Queensland for homesteading and who established vast herds of cattle that could be driven (droved) from place to place over hundreds of square miles, following the patchy, rain-stimulated fodder and the often-ephemeral waterholes - these were the Queensland Overlanders.

The first of the pioneering overlanders were Patsy Durack and his siblings, his wife Mary Costello Durack, their sons and daughters, and brother-in-law John Costello and his family. Between 1867 and 1877 they occupied and claimed 17,000 square miles of land straddling much of the western Channel Country of central Queensland, from the Diamantina River east across and beyond Cooper Creek. The history of their occupation is eloquently described in the book "Kings in Grass Castles" by Patsy Durack's granddaughter, Mary Durack (1990).

On an even larger scale and about a decade or two later Sidney Kidman and his brother Sackville began to acquire much of the land between the Durack's and the Northern Territory border astride the remaining portions of the Channel Country, including the Georgina River, part of the Diamantina drainage and the lower Cooper Creek. It was their intent to develop a chain of stations that would extend from the subtropical north, down along the entire length of the Lake Eyre Basin, ending near the trucking depot at Farina in South Australia. Sackville Kidman died in 1899, but Sidney carried their dream forward, succeeding to establish the greatest pastoral empire in Australia of 68 stations and more than 100,000 square miles at the time of his death in 1935. The history of Sir Sidney Kidman and his holdings is vividly described in Jill Bowen's book, "Kidman: the Forgotten King" (1987).

Both Durack and Kidman soon became aware of the climatic vagaries of central Australia and attempted to compensate for periodic drought and uncertainty with huge land holdings. They knew that even in spite of drought, monsoonal rains in the north would cause the three rivers (Georgina, Diamantina, Cooper Creek) "to come down" in flood periodically, replenishing waterholes and yielding rich grasses in the flood plain. By continually shifting their cattle, they could generally stay ahead of any but the worst drought, providing the cattle with adequate water and forage as they were driven overland to market. However, on open range the cattle and the drovers would concentrate on the regions of richest forage, likely reducing them to stubble before moving on. These preferred regions of highest productivity would also have been those preferred by wildlife, whose sources of food and water could rapidly disappear down the mouths of livestock, and whose nests and burrows could be trampled under hooves. In a land whose usual condition is drought, the fertile flood plains could quickly become degraded, with little chance of regeneration, except in the best of years, drastically altering the continuing productivity of the landscape (Friedel, et al. 1990). The rather precarious, nearly nomadic existence of these earliest pastoralists was expressed aptly in the words of Patsy Durack in 1878, "Cattle kings ye call us, then we are kings in grass castles that may be blown away upon a puff of wind..." (Durack 1990). Unwittingly, these pioneers were erecting grass castles whose ephemeral existences applied not only to themselves and their cattle, but also to the native plants and animals of the region.

It is my belief, based upon observation of portions of the Centre, upon reading historical works, and my own intuition, that while the overlanding days may have started the trend toward habitat degradation and species elimination, that pioneering era was less harmful to the landscape and wildlife than that which was to exist later and continuing to the present. These earlier holders of enormous properties would have concentrated their cattle on the major waterholes and flood plains, particularly in times of drought, leaving at least a moderate probability of surviving smaller islands of refugia here and there. Following droughts these refugia could serve as reservoirs of persistence, contributing to at least some recolonization of nearby recovering, formerly decimated areas.

Since those early pastoral days, however, state and federal restrictions on the size of land holdings have squeezed downward the sizes of stations and increased the numbers of ownerships. In addition, the discovery in the 1880's of a vast series of underground aquifers under much of central Australia led to the rapid development in the 1900's of artesian bores, bringing water to the surface by drilling. Concomitant with this was expansion into the interior of railways for the transport of cattle, to be replaced later by transport trucks, and an incessant subdivision of the land by barbed-wire fencing. All of this had the potential of isolating livestock into relatively small paddocks, resulting in the possibility of a much more uniform and dense distribution of grazing animals across the entire landscape. The likelihood that persistent refugia of native plants and animals would survive this scheme was greatly reduced, even though the bores provided much more broadly distributed and reliable sources of drinking water for those native animals that could access it. With the availability of stock water now guaranteed by the presence of bores, fodder became the key issue. During and after good rainfall periods the temptation was to employ a high utilization/quick turnoff system of management, in which maximal numbers of livestock were allowed to graze a paddock until it became depleted, then moving them out before they lost condition (Friedel, et al. 1990). While maximizing short-term profits during good years and seasons, this strategy results in progressive rangeland degradation, and long term loss of profit as well as habitat deterioration and wildlife loss. At the other end of the management spectrum is the strategy of drought resistance which relies on lower stocking rates and higher per animal production, with a relatively constant income whether seasons are good or bad; this strategy promotes the potential of continuing livestock and wildlife coexistence through prudent resource use and conservation measures. Unfortunately, unrealistically high stocking rates have been required or condoned by government, and government drought relief schemes have subsidized and even rewarded overstocking during hard times, thereby tipping the management balance toward the high utilization/quick turnoff strategy, resulting in continuing massive habitat and rangeland degradation (Friedel, et al. 1990). The view in the outback persists that "the country always comes back when it rains", but that view is not borne out by fact, leaving instead a progressively bleaker scene. These still current practices have resulted in massive habitat degradation and destruction, escalating the rates and probabilities of plant and animal extinctions in central Australia, characteristic of the 20th and now the 21st century.

My personal experiences with these processes have arisen through my several visits to Sandringham and Glengyle stations (particularly the former) in southwestern Queensland, both Kidman holdings and previously important links in Sidney Kidman's chain of stations. Sir Sidney Kidman purchased the two stations from previous owners in 1903, at which time Sandringham was 3,068 square miles and Glengyle, 6,183 square miles; these stations have been reduced to their present areas of 1,908 (Sandringham) and 2,123 (Glengyle) square miles (Bowen 1991), and are now subdivided into many subunits by fencing. The earliest record I can find of Sandringham Station as an entity and Kidman's awareness of it was a statement that he had bought cattle there in 1884 from an earlier owner (Bowen 1991), so its existence as a cattle station exceeds 100 years.

I have described my several visits to Sandringham Station, commencing in 1966, in earlier chapters, and have commented briefly on the three managers I have come to know there successively, Reg Arthur, PeeWee

Clarke, and Geoff Schrader. These gentlemen are skilled stockmen whose jobs are to produce beef, continuing the tradition of Sir Sidney, as epitomized in a verse from the poem "Kidman's Boys" by Les Daley (see Bowen 1987), commemorating the occasion of Kidman's 75th birthday in 1932:

For these are the men of the stations
Who ride 'neath the Northern stars' light,
Where the saltbush blows and the mulga grows
And men must be men in the fight.
Where they're not yarded up by the tramlines
And no boundary of brick wall annoys,
A thousand mile ride they take in their stride,
'Tis the day's work for Kidman's boys.

They and their wives are both tough and gentle people, who love the lands they manage, and are proud to be a part of their country's history. Chronologically Reg, PeeWee and Geoff are about half a generation apart, with Reg, now retired, being the elder. What I have found with each of these gentlemen is that, as younger managers, they are so preoccupied with establishing their management as a profitable enterprise for the Kidman Co., that they have little or no time to think beyond the very substantial day to day business of cattle, weather, hired hands and station operation. It is not until they become older that they begin to express a fondness and a concern for the natural elements that constitute their station, and indeed, upon which its livelihood depends. These are fundamental, human progressions parallel to those which we all experience, and which I have nearly invariably witnessed in academics. Therefore, Geoff Schrader puts all of his time, thought and effort into producing marketable cattle, as did PeeWee Clarke when I first met him in 1974. Reg Arthur was more senior when I first met him in 1966 and was more philosophically reflective about the landscape and its dependent fauna. PeeWee Clarke, in our recent visit, had joined Reg in a shared concern for environmental degradation and faunal disappearance, but they have been unable to grapple with the underlying causes and their resolutions.

During the period of my sporadic visits to Sandringham I believe that I have personally witnessed the actual or near local extinctions of two of the intermediate-sized, carnivorous/insectivorous dasyurid marsupials, the Kultarr (*Antechinomys laniger*) and the Kowari (*Dasyuroides byrnei*). Over the same time span (1966 to present) Pat Woolley from La Trobe University, in all likelihood has documented the local extinction on Sandringham of another, similar-sized dasyurid, the Mulgara (*Dasycercus cristicauda*) (Woolley 1989; personal communications). On Sandringham, the Mulgara is a sand hill specialist, and it is considered an endangered species (Kennedy 1990). In our most recent trip to Sandringham in 1992, Pip

Masters, who is very familiar with the Mulgara, was unable to locate any at the sites previously identified by Woolley, or at other similar sand hill sites on Sandringham. It is interesting to note that these three dasyurid marsupials share several characteristics that likely contribute to their vulnerability, although just how is unclear. They are all smallish in size, rat-size to weasel-size, they are insectivorous/carnivorous, and they occur only in arid settings. They differ in habitat preference, however, in that the Mulgara lives in sand hills, while the Kultarr and Kowari are inhabitants of Gibber plain.

Our collecting efforts for mammals at Sandringham were concentrated on clay pan and gibber habitats, rather than on sand hills, and would have missed the presence of the Mulgara, had they been extant, especially in 1966 and 1974. The results of our efforts, employing approximately equivalent techniques and intensities are summarized in Table 2, showing the disappearance of the Kultarr and the Kowari. Although the abundance of rodent species appears constant through 1992, the Table is somewhat misleading, as our collecting of Fawn Hopping-mice was for a fixed number of experimental specimens, rather than maximal capture success. I can say with complete confidence however, that the Fawn Hopping-mouse has experienced declines over this period in both abundance and distribution, and, while their populations are at present reasonably healthy, they are threatened with progressive future declines and possible extirpation especially by introduced predators. They, too, deserve protection.

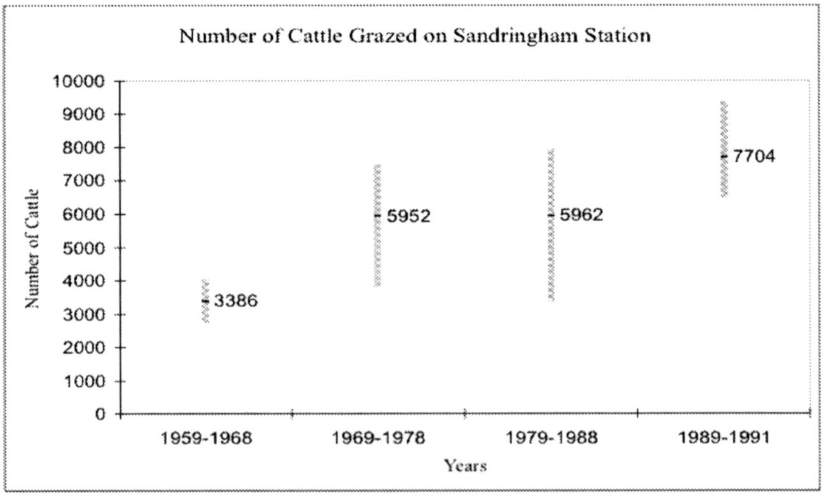

Figure 19. Numbers of cattle on Sandringham Station, southwestern Queensland (stocking figures) since 1959 (Woolley 1989; George Birchmore, personal communication). The bars enclose the ranges of annual counts, and the horizontal lines and numbers represent the means for the respective periods.

Habitat degradation resulting from cattle grazing no doubt is a contributing factor to the demise of these species, for Sandringham is under rather intensive, and perhaps increasing grazing pressure as stocking records show (fig. 19). However, grazing alone is certainly not the only culprit. I am quite certain that two introduced predators, the domestic cat and the European fox, figure prominently in this race to extinction, an opinion shared also by Tim Low in his recent (2002) book, "Feral Future". Feral cats and/or their tracks were seen widely spread over the station, and we witnessed European foxes actively foraging in prime hopping-mouse habitat. In 1992, we saw the tracks of a cat near the Seventeen-mile waterhole and the Montara sand hill, one of Pat Woolley's earlier Mulgara sites. Unwanted young cats frequently wander from the station homesteads to join the feral population, and station people occasionally "dump" entire litters in the bush, rather than killing the unwanted ones. In 1974 at Sandringham Terry Dawson and I persuaded Diane Clark to let us drown 11 of 13 kittens from two litters that she otherwise was going to release in the bush. She hadn't really thought of the havoc they might have played with the native fauna, but seemed genuinely concerned when we addressed this issue. I suspect that habitat degradation through heavy grazing in fenced paddocks places sensitive mammalian species at risk and that predation by and competition with introduced mammalian predators tips the balance toward extinction.

It is interesting to speculate further that the naturalized large predator, the Dingo, may play a role here in its increasing absence. Dingoes have been under intensive attack by graziers throughout the Centre, as depredators of livestock, both calves and sheep, and Terry Dawson and I in 1974 witnessed their effectiveness as predators in mortally wounding a calf not far from Sandringham station. Dingoes are actively hunted, trapped and poisoned in Queensland today, and we saw very few signs of them on our most recent visit to Sandringham. Their presence in Australia for at least 3,000 years means that they have coexisted for a long while with the native mammalian fauna, whose rapid declines have coincided with the later European settlement. Studies by Newsome, et al. (1984) on Australia's east coast document that Dingoes prefer large (supplementary prey) and medium-sized (staple prey) mammalian prey, with small mammals in the size classes of those in Table 2 from Sandringham Station designated as opportune prey taken only when abundant. Their (ibid.) study also shows that Dingoes will eat both European foxes and domestic cats, although they were considered only minor dietary items in their study. Foxes and cats would fall into their category of medium-sized mammals or staple prey. I postulate that the Dingo, when present in normal numbers (i.e., not decimated by humans), reduces by predation the numbers of foxes and feral cats to lower levels consistent with continuing existences of small native mammals. This postulate would also predict that predation pressure on these small native mammals by the Dingo is insignificant. Alan Newsome, a specialist on predator-prey relations in Australia, agreed that such a postulate made sense (personal communication). Thus, the Dingo's presence in rangelands is enigmatic, where it is viewed by pastoralists as a killer of livestock, and by conservationists as a protector of native animals.

The scenario that arises from our observations suggests that modern grazing practices, where cattle or sheep are overstocked in fenced paddocks and supplied with bore water, gradually alter native habitats to levels of marginality or worse. The low mobility (i.e., sedentary) of specialist small native mammals progressively restricts them to these increasingly marginal habitat islands. In the marginal habitat islands small mammals are particularly vulnerable to predation, especially to that imposed by effective, voracious introduced predators, such as feral cats and foxes. In the presence of the Dingo, cats and foxes may themselves become prey, thereby reducing the predation pressure on small native mammals. If Dingoes are eradicated as livestock predators, the cats and foxes may become established and decimate populations of small native mammals, resulting in local or widespread extinctions. Dingo control without simultaneous control of introduced predators, such as cats and foxes, will, according to this scenario, result in continuing extinctions such as we have apparently witnessed at Sandringham Station.

Parts of this scenario are consistent with one proposed recently by Steve Morton (1990), then a wildlife specialist with CSIRO Centre for Arid Zone Research, Alice Springs, and now Group Executive, CSIRO's Sustainable Energy & Environment Unit, Canberra. Steve agrees that the driving force in the high rate of extinction of small and medium-sized central Australian mammals is habitat degradation and destruction concomitant with European settlement, and the arrival of herbivorous livestock and rabbits (introduced herbivores), and the pressure of introduced predators. However, Steve's scenario lays the principal blame for habitat degradation on rabbits, which may be true throughout portions of central Australia, but not Sandringham Station where I have never seen a rabbit, and where certainly they have never occurred in more than insignificant numbers thus far. The only introduced herbivores of consequence on Sandringham Station are cattle.

What is in store for the future for places like Sandringham Station? That is a very difficult question to answer, with social, economic and political repercussions, for it is clear to me at least that the prevailing central Australian pastoral tradition, and native plant and animal life are on a collision course. Hopefully, what we have witnessed at Sandringham are only localized extinctions, and isolated pockets of healthy habitat with these native mammal species may persist. However, unless these pockets are identified and steps taken to protect and preserve them, they remain vulnerable to the continuing pressures that lead to extinction. To divert the course of events from collision to coexistence, and continuing survival would require substantial changes in land management practices, altering the lifestyles of pastoralists who have over a hundred years of tradition. This is not easy, particularly given the time of probably no more than a few human generations to resolve the problem. Although I have not discussed these issues directly with pastoralists it was clear that they are apprehensive, for the managers of Sandringham Station on our last visit (1992), the Schraders, were fearful that the local stations in that area might be targets for National Parks status.

Perhaps there is still time for an amicable resolution of this conflict between the pastoralist tradition and wildlife, since a possible model for enlightened land-management is presently being applied in the Centre at Uluru National Park in the Northern Territory. Unlike the large cattle holdings of the Channel Country, Uluru has not been subject to commercial grazing pressure, but it has suffered severe habitat degradation from populations of two introduced, now feral herbivores, the Camel and the European rabbit. Like Sandringham, though, populations of two introduced predators are well established at Uluru, the domestic cat and the European fox (Reid, et al. 1993). During the period of European settlement, these interacting competitive and predatory pressures have contributed in the region around Uluru to the decline of small and medium-sized native mammals. The original 46 species of terrestrial mammals has been reduced to only 22 species currently (Creagh 1992), a loss of more than one-half of the native species. Uluru National Park, which includes both Uluru (Ayers Rock) and Kata Tjuta (The Olgas), was owned by the Australian Federal Government and, during the latter half of the 20th century, it rapidly became an internationally recognized tourist attraction. In 1985, the National Park was returned to its traditional Aboriginal owners, the Anangu who, in turn, leased the Park back to the Australian National Parks and Wildlife Service. Part of the lease agreement required the establishment of the Uluru-Kata Tjuta Board of Management and a Plan of Management in which personnel equally from National Parks and Wildlife and from the Anangu Community jointly manage the Park. The Plan employs traditional Aboriginal land-management skills and philosophies of caring for the land as codified in their Dreaming, together with scientific and experimental skills of land-management developed by the National Parks and Wildlife.

This joint management is still in the exploratory stages, but it appears to be working, and its transition has achieved by far the greatest degree of harmony between formerly conflicting land ownership and land use, and for the seeming benefit of wildlife and the welfare of the Aboriginal people.

The situation at Sandringham and Queensland's Channel Country certainly differs from that of Uluru, but there are many obvious parallels, too. It differs in that much of the land is under private ownership, and there are far fewer Aboriginals, scattered across the country as station hands, or confined to reserves in small towns, such as Birdsville. The persons who possess the greatest knowledge about the country, at least biogeographically, are the station people themselves, particularly the managers and their families. They actually or potentially possess the skills and experience for the soundest land management, although they lack the scientific expertise. It seems to me that a mutually productive arrangement could be struck between the Australian National Parks and Wildlife Service and major land holding companies or even individual stations. They could initiate programs of joint land management in which commercial grazing would employ a drought-resistance strategy with low sustainable stocking rates, and with equal emphasis placed on native habitat and wildlife preservation and restoration. The traditional lifestyles of the pastoralists could be retained and their wealth of knowledge and experience fruitfully employed as habitat and wildlife managers, in addition to managers of livestock. Certainly not to be excluded from such a scheme are local Aboriginals, where present, as their knowledge of land management transcends scores if not hundreds of generations, as passed through oral tradition. An additional layer of carefully controlled tourism could be superimposed on this scheme, to generate compensatory income.

Admittedly, this scheme would require both monetary sacrifice and investment in the short term, and willingness both of landowners and their employees, and state and federal governments, and a level of cooperation that rarely has been seen. But the long term stakes are great, for they include the probable extinction or salvation of elements of a truly unique and magnificent fauna and flora, some of which have disappeared before my eyes, and others which appear to be rapidly following. Another element of an anthropogenic nature that I have not addressed but must mention as having potentially deleterious impacts of unknown magnitude and consequence on this region is that related to global atmospheric interference, particularly greenhouse gases and their relationship to global climatic warming, and chlorofluorocarbons and their role in depletion of the protective ozone layer in the upper atmosphere. There are many impediments standing in the way of safeguarding this unique ecosystem from further degradation and extinction, but in my estimation, it is not too late. If the status quo persists, it soon will be too late for many species, and what will be left for meanderings then? We must ensure that the future provides opportunities for additional meanderings, and not just memories.

I hope that the preceding observations, descriptions, and attempts at syntheses will have provided the reader with a clearer picture of the natural wonders of the land down under, and particularly of a segment of its interior and the unique animals and people that occupy it. In addition, I am hopeful that somehow I have conveyed the necessity, importance, excitement, and sheer fun of conducting fieldwork as a prelude to laboratory-borne experimental work in environmental biology. Each is equally important, and neither should stand alone. Finally, I hope that I have managed to make these writings intriguing to the broad audience for which it is intended, and that includes casual and professional naturalists of all ages, levels of expertise, and persuasions. Moreover, please, always preserve room for meanderings, whether they are mental, emotional, physical, or any combination thereof, for they will help to enrich and broaden your horizons.

SUGGESTED FURTHER READING

Bowen, J. 1987. Kidman: the Forgotten King. North Ryde: Angus & Robertson.

Durack, M. 1990. Kings in Grass Castles. Moorebank: Corgi Books

Flannery, T. 1995. The Future Eaters. New York: George Braziller.

Low, T. 2002. Feral Future: the Untold Story of Australia's Exotic Invaders. Chicago: University of Chicago Press.

Mulvaney, J. and J. Kamminga. 1999. Prehistory of Australia. St. Washington: Smithsonian Institution Press.

POSTSCRIPT

Almost precisely 10 years after our final visit to Sandringham Station, Barbara, Ian (then 18, and in college), and I returned to central Australia in June 2002 (fig. 20), to look at the deserts of Western Australia, and to contrast them with those of the Lake Eyre Basin.

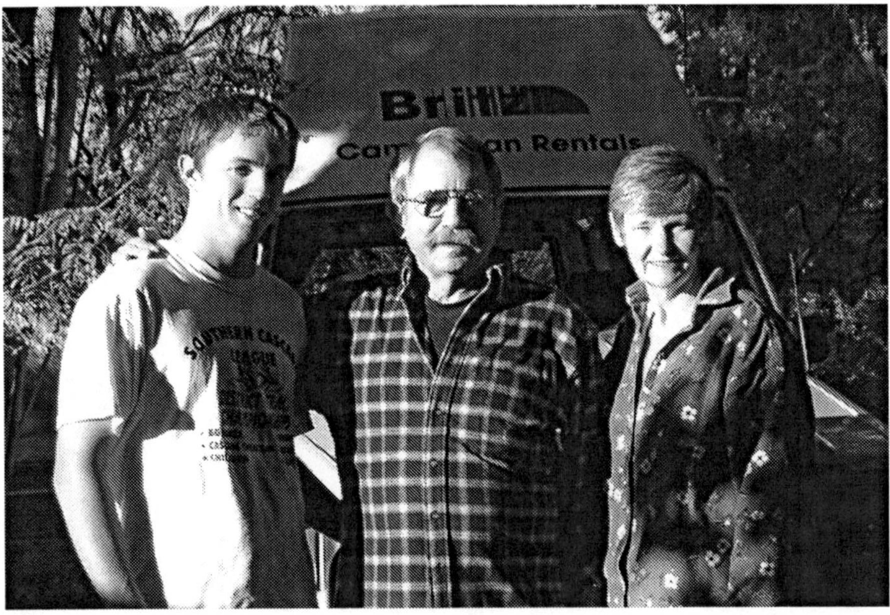

Figure 20. The MacMillens: Ian, Dick, and Barb about to depart from Alice Springs, 2002. Photo by Lynn Day.

On 9 June 2002 we departed Uluru (Ayers Rock), Northern Territory in a rental 4-wheel drive Toyota "Bushcamper", to travel in a southwesterly direction toward Perth, Western Australia, on a dirt track generously labeled

"The Great Central Road" (see map 1). This road traverses parts of the Gibson Desert and the Great Victoria Desert, passing through several Aboriginal Lands for which we held transit permits, and continues as a dirt track between Uluru and the W.A. town of Laverton for a distance of 1127 km (704 mi). From Laverton to Perth (another 917 km = 578 mi) the roads are all sealed (paved). As it was commencing winter, and we drove only by day and did no spotlighting or trapping at night, only a modest amount of wildlife was seen, the most notable of which were Dingoes and feral Camels, both human introductions. In addition, several parrot species were seen, along with Crested Pigeons, crows, Emus, a single Mulga Snake, and the occasional kangaroo. The most interesting feature that contrasted with the Lake Eyre Basin was the nearly complete absence of major creek and river channels, with their majestic River Red Gums and Ghost Gums that loom over their usually dry, sandy courses that we so loved in the Channel Country. Between Uluru and the Northern Territory/Western Australia border, near the Aboriginal community of Docker River, N.T., we left the last major water courses behind, the Hull River and the Docker River. These represent the westernmost rivers we crossed that, at flood stage, send their waters toward the Lake Eyre Basin. Shortly after entering Western Australia and the Gibson Desert the land became progressively featureless, dominated by frequent sand hills that were pretty much stabilized with vegetation, and the occasional rock outcrop. Rather than major waterways that dissect the landscape as in the Channel Country, the land appeared to be divided into countless, small internal drainage systems, where, following the infrequent rain, water ran for short distances and then collected in the low points, forming numerous clay pans between the sand hills. However, there appeared to be little sign of the kind of expansive flooding that characterizes the Channel Country, where there it must be a major evolutionary force.

Although these Western Australian deserts also get much of their rain as unpredictable summer monsoonal events, the apparent rarity of accompanying flooding along extensive waterways classified them in my book as more typical lower latitude deserts, somewhat comparable to North America's Sonoran Desert. Admittedly, this is a snapshot view of a very complex ecosystem, but the contrasts were conspicuous, and the Channel Country will remain in my mind a world of unique ecological challenge for its biota, for the hardy people who live there, and for those who meander through it.

LITERATURE CITED

Aitken, P. 1979. The Status of Endangered Australian Wombats, Bandicoots and the Marsupial Mole. In The Status of Endangered Australian Wildlife, edited by M.J. Tyler, 61-65. Proc. Adelaide: Centenary Symposium of the Royal Zoological Society of South Australia.

Allan, R.J. 1990. Climate. In Natural History of the North East Deserts, edited by M.J. Tyler, C.R. Twidale, M. Davies and C.B. Wells, 81-84. Adelaide: Royal Society of South Australia.

Anonymous. 1861. The Burke and Wills Exploring Expedition: An Account of the Crossing of the Continent of Australia, from Cooper's Creek to Carpentaria, with Biographical Sketches of Robert O'Hara Burke and William John Wills. Wilson and Mackinnon, Melbourne. (Facsimile Edition of 1963, Adelaide: Libraries Board of South Australia).

Anonymous. 1991. Birdsville, 1991 Guide.

Aschoff, J. and H. Pohl. 1970. Rhythmic Variations in Energy Metabolism. Federation Proceedings 29: 1541-1552.

Barker, W.R. and P.J.M. Greenslade. Evolution of the Flora and Fauna of Arid Australia. Frewville: Peacock Publications.

Barlow, N. (ed). 1933. Charles Darwin's Diary of the Voyage of H.M.S. "Beagle". Cambridge: Cambridge University Press.

Bartholomew, G.A. 1972. The Water Economy of Seed-eating Birds That Survive Without Drinking Water. Proceedings of the XV International Ornithological Congress: 237-254.

Bartholomew, G.A. and V.A. Tucker. 1963. Control of Changes in Body Temperature, Metabolism, and Circulation by the Agamid Lizard, *Amphibolurus barbatus*. Physiological Zoology 36: 199-218.

Baudinette, R.V. 1982. The Energetics of Locomotion in Dasyurid Marsupials. In Carnivorous Marsupials, Vol. I, edited by M. Archer, 261-265. Mosman: Royal Zoological Society of New South Wales.

Baverstock, P.R. 1982. Adaptations and Evolution of the Mammals of Arid Australia. In Evolution of the Flora and Fauna of Arid Australia, edited by W.R. Barker and P.J. Greenslade, 175-178. Frewville: Peacock Publications.

Beckler, H. 1993. A Journey to Cooper's Creek. Carlton, Melbourne: Melbourne University Press.

Berger, P.J., N.C. Negus and P.D. Gardner. 1981. Chemical Triggering of Reproduction in *Microtus montanus*. Science 214: 69-70.

Berra, T.M. 1998. A Natural History of Australia. San Diego: Academic Press.

Biewener, A.A. and R.V. Baudinette. 1995. In Vivo Muscle Force and Elastic Energy Storage During Steady-speed Hopping of Tammar Wallabies (*Macropus eugenii*). Journal of Experimental Biology 198: 1829-1841.

Bishop, J.A. 1963. The Australian Freshwater Crabs of the Family Potamonidae (Crustacea: Decapoda). Australian Journal of Marine and Freshwater Research 14: 218-238.

Blakers, M., S.J.J.F. Davies, and P.N. Reilly. 1985. The Atlas of Australian Birds. Melbourne: Melbourne University Press.

Bowen, J. 1991. Kidman: The Forgotten King. North Ryde: Angus and Robertson.

Bowler, J.M. 1982. Aridity in the Late Tertiary and Quaternary of Australia. In Evolution of the Flora and Fauna of Arid Australia, edited by W.R. Barker and P.J.M. Greenslade, 35-45. Frewville: Peacock Publications.

Bowler, J.M., H. Johnston, J.M. Olley, J.R. Prescott, R.G. Roberts, W. Shawcross, and N. Spooner. 2003. New Ages for Human Occupation and Climatic Change at Lake Mungo, Australia. Nature 421: 837-840.

Bowler, J.M., R. Jones, H. Allen, and A.G. Thorne. 1970. Pleistocene Human Remains from Australia: a Living Site and Human Cremation from Lake Mungo, Western New South Wales. World Archaeology 2: 39-60.

Bowler, J.M., and A.G, Thorne, 1976. Human Remains from Lake Mungo. Discovery and Excavation of Lake Mungo III. In The Origins of the Australians edited by R.L. Kirk and A.G. Thorne, 127-138. Canberra: AIAS.

Breed, W.G. 1982. Control of Mammalian and Avian Reproduction in the Australian Arid Zone, With Special Reference to Rodents. In Evolution of the Flora and Fauna of Arid Australia, edited by W.R. Barker and P.J.M. Greenslade, 185-190. Frewville: Peacock Publications.

Brody, S. 1945. Bioenergetics and Growth. New York: Reinhold.

Cade, T.J., and J.A. Dybas, Jr. 1962. Water Economy of the Budgerygah. Auk 79: 345-364.

Cade, T.J., C.A. Tobin, and A. Gold. 1965. Water Economy and Metabolism of Two Estrildine Finches. Physiological Zoology 38: 9-33.

Cardillo, M., G.M. Mace, K.E. Jones, J. Bielby, O.R.P. Bininda-Emonds, W. Sechrest, C.D.L. Orme, and A. Purvis. 2005. Multiple Causes of High Extinction Risk in Large Mammal Species. Science 309: 1239-1241.

Christidis, L., R. Schodde, D.D. Shaw, and S.F. Maynes. 1991. Relationships Among the Australo-Papuan Parrots, Lorikeets, and Cockatoos (Aves:Psittaciformes): Protein Evidence. Condor 93: 302-317.

Cogger, H.C., and E.E. Cameron (eds). 1984. Arid Australia. Sydney: Australian Museum.

Collins, B.G., and S.D. Bradshaw. 1973. Studies on the Metabolism, Thermoregulation and Evaporative Water Losses of Two Species of Australian Rats, *Rattus villosissimus* and *Rattus rattus*. Physiological Zoology 46: 1-21.

Corbett, L.K. 1985. Morphological Comparisons of Australian and Thai Dingoes: a Reappraisal of Dingo Status, Distribution and Ancestry. In Ecology of the Wet-Dry Tropics, edited by M.G. Ridpath and L.K. Corbett, 277-291. Melbourne: Proceedings of the Ecological Society of Australia 13.

_______. 1995. Dingo. In Mammals of Australia, edited by R. Strahan, 696-698. Washington, D.C.: Smithsonian Institution Press.

Craig, J.L. and R.E. MacMillen. 1985. Honeyeater Ecology: an Introduction. New Zealand Journal of Zoology 12: 565-568.

Creagh, C. 1992. Looking After the Land at Uluru. Ecos 71: 6-13.

Davies, S.J.J.F. 1976. The Natural History of the Emu in Comparison With That of Other Ratites. Proceedings of the XVI International Ornithological Congress, 109-120. Canberra: Australian Academy of Science.

Davies, S.J.J.F., M.W.R. Beck and J.P. Kruiskamp. 1971. The Results of Banding 154 Emus in Western Australia. C.S.I.R.O. Wildlife Research 16: 77-79.

Dawson, T.J. 1998. Kangaroos: The Biology of the Largest Marsupial. Ithaca: Cornell University Press.

Dawson, T.J. and A.J. Hulbert. 1970. Standard Metabolism, Body Temperature and Surface Area of Australian Marsupials. American Journal of Physiology 218: 1233-1238.

Dawson, T.J. and C.R. Taylor. 1973. Energetic Cost of Locomotion in Kangaroos. Nature 246: 313-314.

Dawson, T.J., R.M. Herd and E. Skadhauge. 1983. Water Turnover and Body Water Distribution in a Large Arid-zone Bird, the Emu, *Dromaius novaehollandiae*. Journal of Comparative Physiology B153: 235-240.

Dawson, T.J., D. Read, E.M. Russell and R.M. Herd. 1984. Seasonal Variation in Daily Activity Patterns, Water Relations and Diet of Emus. Emu 84: 93-102.

Dawson, W.R., and A.F. Bennett. 1973. Roles of Metabolic Level and Temperature Regulation in the Adjustment of Western Plumed Pigeons (*Lophophaps ferruginea*) to Desert Conditions. Comparative Biochemistry and Physiology 44: 249-269.

Dayton, L. 2003. Tracing the Road Down Under. Science 302: 555.

_______. 2003. On the Trail of the First Dingo. Science 302: 555-556.

Denny, M.J.S. 1982. Adaptations of the Red Kangaroo and Euro (Macropodidae) to Aridity. In Evolution of the Flora and Fauna of Arid Australia, edited by W.R. Barker and P.J.M. Greenslade, 179-183. Frewville: Peacock Publications.

Durack, M. 1990. Kings in Grass Castles. Moorebank: Corgi Books.

Edwards, P. (ed.). 1999. The Journals of Captain Cook. New York: Penguin Books.

Eriksen, C.H. and D. Belk. 1999. Fairy Shrimp. Eureka: Mad River Press.

Fisher, C.D., E. Lindgren and W.R. Dawson, 1972. Drinking Patterns and Behavior of Australian Desert Birds in Relation to Their Ecology and Abundance. Condor 74: 111-136.

Flannery, T.F. 1995. The Future Eaters. New York: George Brazilier, Inc.

_______. 1999. Debating Extinction. Science 283: 182-183.

Flood, J. 1989. Archaeology of the Dreamtime. Sydney: Collins Publishers.

Ford, H.A. 1979. Interspecific Competition in Australian Honeyeaters - Depletion of Common Resources. Australian Journal of Ecology 4: 145-164.

Ford, H.A., and D.C. Paton. 1977. The Comparative Ecology of Ten Species of Honeyeaters in South Australia. Australian Journal of Ecology 2: 399-407.

Forster, P. and S. Matsumura. 2005. Did Early Humans Go North or South? Science 308: 965-966.

Friedel, M.H., B.D. Foran and D. M. Stafford-Smith. 1990. Where the Creeks Run Dry or Ten Feet High: Pastoral Management in Arid Australia. Proceedings of the Ecological Society of Australia 16: 185-194.

Friend, J.H. and D.B. Guralnik (eds.). 1960. Webster's New World Dictionary. Cleveland and New York: World Publishing Co.

Frith, H.J. 1982. Pigeons and Doves of Australia. Adelaide: Rigby Publishers.

Geiser, F. 2004. The Role of Torpor in the Life of Australian Arid Zone Mammals. Australian Mammalogy 26: 125-134.

Glover, C.J.M. 1990. Fishes. In Natural History of the North East Deserts, edited by M.J. Tyler, C.R. Twidale, M. Davies, and C.B. Wells, 189-197. Northfield: Royal Society of South Australia.

Gordon, M.S., G.A. Bartholomew, A.D. Grinnell, C.B. Jorgensen, and F.N. White. 1977. Animal Physiology: Principles and Adaptations. New York: Macmillan Publishing Company.

Greenaway, P. 1984. Survival Strategies in Desert Crabs. In Arid Australia, edited by H.G. Cogger and E.E. Cameron, 145-152. Sydney: Australian Museum.

Greenaway, P., and R.E. MacMillen. 1978. Salt and Water Balance in the Terrestrial Phase of the Inland Crab, *Holthuisana (Austrothelphusa) transversa* Martens (Parathelphusoidea: Sundathelphusidae). Physiological Zoology 51: 217-229.

Hall, N., R. Boden, C.S. Christian, R. Condon, F. Dale, A. Hart, J. Leigh, J. Marshall, A. McArthur, V. Russell and J. Turnbull. 1972. The Use of Trees and Shrubs in the Dry Country of Australia. Canberra: Australian Government Publishing Service.

Heatwole, H. 1987. Major Components and Distributions of the Terrestrial Fauna. In Fauna of Australia, edited by G.R. Dyne and D.W. Walton, Ch.5. General Articles, Vol. 1a. Canberra: Australian Government Publishing Service.

Hinds, D.S. and R.E. MacMillen. 1984. Energy Scaling in Marsupials and Eutherians. Science 225: 335-337.

Johnson, B.J., G.H. Miller, M.L. Fogel, J.W. Magee, M.K. Gagan, and A.R. Chivas. 1999. 65,000 Years of Vegetation Change in Central Australia and the Australian Summer Monsoon. Science 284: 1150-1152.

Jones, R. 1969. Fire-stick Farming. Australian Natural History 16: 224-228.

Kalma, J.D. and J.R. McAlpine. 1979. Climate and Man in the Centre. In Man in the Centre, edited by G. Crook, 46-69. Melbourne: CSIRO.

Kemper, C.M. 1990. Mammals. In Natural History of the North East Deserts, edited by M.J. Tyler, C.R. Twidale, M. Davies, and C.B. Wells, 161-167. Northfield: Royal Society of South Australia.

Kennedy, M. 1990. Australia's Endangered Species. Brookvale: Simon and Schuster Australia.

Kennedy, M. and R. Burton. 1986. A Threatened Species Conservation Strategy for Australia. Manly: Ecofund Australia.

Kimber, R.G. 1986. Man from Arltunga. Carlisle: Hesperian Press.

Kleiber, M. 1932. Body Size and Metabolism. Hilgardia 6: 315-353.

Lee, A.K. and A. Cockburn. 1985. Evolutionary Ecology of Marsupials. Cambridge: Cambridge University Press.

Lee, A.K. and E.H. Mercer. 1967. Cocoon Surrounding Desert-dwelling Frogs. Science 157: 87-88.

Lee, A.K., P. Wooley, and R.W. Braithwaite. 1982. Life History Strategies of Dasyurid Marsupials. In Carnivorous Marsupials, Vol.I, edited by M. Archer, 1-11. Mosman: Royal Zoological Society of New South Wales.

Lindgren, E. 1973. Studies in the Ecology and Physiology of Three Species of Grass-parrots. Ph.D. Thesis. University of Western Australia.

Low, T. 2002. Feral Future: the Untold Story of Australia's Exotic Invaders. Chicago: University of Chicago Press.

Mackellar, Dorothea. 1911. "My Country". In The Closed Door, and Other Verses. Melbourne: Australasian Authors' Agency.

MacMillen, R.E. 1972. Water Economy of Nocturnal Desert Rodents. Symposium of the zoological society of London 31: 147-174.

_______. 1981. Nonconformance of Standard Metabolic Rate with Body Mass in Hawaiian Honeycreepers. Oecologia 49: 340-343.

_______. 1985. Energetic Patterns and Lifestyle in the Meliphagidae. New Zealand Journal of Zoology 12: 623-629.

_______. 1990. Water Economy of Granivorous Birds: A Predictive Model. Condor 92: 379-392.

MacMillen, R.E. and R.V. Baudinette. 1993. Water Economy of Granivorous Birds: Australian Parrots. Functional Ecology 7: 704-712.

MacMillen, R.E. and T.J. Dawson. 1986. Energy and Water Metabolism of the Kowari, *Dasyuroides byrnei* (Marsupialia: Dasyuridae), While Resting and Running. Australian Mammalogy 9: 87-95.

MacMillen, R.E. and P. Greenaway. 1978. Adjustments of Energy and Water Metabolism to Drought in an Australian Arid-zone Crab. Physiological Zoology 51: 230-240.

MacMillen, R.E. and A.K. Lee. 1967. Australian Desert Mice: Independence of Exogenous Water. Science 158: 383-385.

_______. 1969. Water Metabolism of Australian Hopping Mice. Comparative Biochemistry and Physiology 28: 493-514.

_______. 1970. Energy Metabolism and Pulmocutaneous Water Loss of Australian Hopping Mice. Comparative Biochemistry and Physiology 35: 355-369.

MacMillen, R.E. and J.E. Nelson. 1969. Bioenergetics and Body Size in Dasyurid Marsupials. American Journal of Physiology 217: 1246-1251.

MacMillen, R.E., M.L. Augee and B.A. Ellis. 1989. Thermal Ecology and Diet of Some Xerophilous Lizards from Western New South Wales. Journal of Arid Environments 16: 193-201.

MacMillen, R.E., R.V. Baudinette and A.K. Lee. 1972. Water Economy and Energy Metabolism of the Sandy Inland Mouse, *Leggadina hermannsburgensis*. Journal of Mammalogy 53: 529-539.

Main, B.Y. 1982. Adaptations to Arid Habitats by Mygalomorph Spiders. In Evolution of the Flora and Fauna of Arid Australia, edited by W.R. Barker and P.J.M. Greenslade, 273-283. Frewville: Peacock Publications.

Marshall, A.J. and R. Drysdale. 1966. Journey Among Men. Melbourne: Sun Books.

McCullough, D.R. and Y. McCullough. 2000. Kangaroos in Outback Australia. New York: Columbia University Press.

McNight, T. 1976. Friendly Vermin. A Survey of Feral Livestock in Australia. University of California Publications in Geography 21: 1-104.

Miller, A.H. 1963. Desert Adaptations in Birds. Proceedings of the XIII International Ornithological Congress, 666-674.

Miller, G.H., J.W. Magee, B.J. Johnson, M.L. Fogel, N.A. Spooner, M.T. McCulloch, and L.K. Ayliffe. 1999. Pleistocene Extinction of *Genyornis newtoni*: Human Impact on Australian Megafauna. Science 283: 205-208.

Moorehead, A. 1985. Cooper's Creek. Melbourne: Sun Books.

Morton, R.K. and E. Barber. 2004. The Truth and Adventures of Serendipity: A Study in Sociological Semantics and the Sociology of Science. Princeton: Princeton University Press.

Morton, S.R. 1978. Torpor and Nest-sharing in Free-living *Sminthopsis crassicaudata* (Marsupialia) and *Mus musculus* (Rodentia). Journal of Mammalogy 59: 569-575.

_______. 1980. Field and Laboratory Studies of Water Metabolism in *Sminthopsis crassicaudata* (Marsupialia: Dasyuridae). Australian Journal of Zoology 28: 213-227.

_______. 1982a. Dasyurid Marsupials of the Australian Arid Zone: an Ecological Review. In Carnivorous Marsupials. Vol. I, edited by M. Archer, 117-130. Mosman: Royal Zoological Society of New South Wales.

_______. 1982b. Granivory in the Australian Arid Zone: Diversity of Harvester Ants and Structure of Their Communities. In Evolution of the Flora and Fauna of Arid Australia, edited by W.R. Barker and P.J.M. Greenslade, 257-262. Frewville: Peacock Publications.

_______. 1990. The Impact of European Settlement on the Vertebrate Animals of Arid Australia: a Conceptual Model. Proceedings of the Ecological Society of Australia 16: 201-213.

Morton, S.R. and P.H. Davies. 1983. Food of the Zebra Finch (*Poephila guttata*), an Examination of Granivory in Birds of the Australian Arid Zone. Australian Journal of Ecology 8: 235-243.

Morton, S.R., M. Happold, A.K. Lee, and R.E. MacMillen. 1977. The Diet of the Barn Owl, *Tyto alba*, in South-western Queensland. Australian Wildlife Research 4: 91-97.

Mulvaney, J. and J. Kamminga. 1999. Prehistory of Australia. Washington: Smithsonian Institution Press.

Myers, F.R. 1991. Pintupi Country, Pintupi Self. Berkeley: University of California Press.

Newsome, A.E. 1965. Reproduction in Natural Populations of the Red Kangaroo, *Megaleia rufa* (Desmarest), in Central Australia. Australian Journal of Zoology 13: 735-759.

_______. 1966. The Influence of Food on Breeding in the Red Kangaroo in Central Australia. CSIRO Wildlife Research 11: 187-196.

_______. 1980. The Eco-mythology of the Red Kangaroo in Central Australia. Mankind 12: 327-333.

Newsome, A.E., P.C. Catling and L.K. Corbett. 1984. The Feeding Ecology of the Dingo II. Dietary and Numerical Relationships with Fluctuating Prey Populations in South-eastern Australia. Australian Journal of Ecology 8: 345-366.

Newsome, A.E. and L.K. Corbett. 1975. Outbreaks of Rodents in Semi-arid and Arid Australia: Causes, Preventions, and Evolutionary Considerations. In Rodents in Desert Environments, edited by I. Prakash and P.K. Ghosh, 117-153. The Hague: Junk Pubs.

Pianka, E.R. 1986. Ecology and Natural History of Desert Lizards. Princeton: Princeton University Press.

Reid, J.R.W., J.A. Kerle and S.R. Morton. 1993. Uluru Fauna. The Distribution and Abundance of Vertebrate Fauna of Uluru (Ayers Rock-Mount Olga) National Park, N.T. Australian National Parks and Wildlife Service Series - Kowari 4.

Roberts, R.G., R. Jones and M.A. Smith. 1990. Thermoluminescence of a 50,000-year-old Human Occupation Site in Northern Australia. Nature 345: 153-156.

Roberts, R.G., T.F. Flannery, L.K. Ayliffe, H. Yoshida, J.M. Olley, G.J. Prideaux, G.M. Laslett, A. Baynes, M.A. Smith, R. Jones and B.L. Smith. 2001. New Age for the Last Australian Megafauna: Continent-wide Extinction About 46,000 Years Ago. Science 292: 1888-1892.

Rosenzweig, M.L. 1968. Net Primary Productivity of Terrestrial Communities: Prediction from Climatological Data. American Naturalist 102: 67-74.

Schmidt-Nielsen, K., and A.E. Newsome. 1962. Water Balance in the Mulgara (*Dasycercus cristicauda*), a Carnivorous Desert Marsupial. Australian Journal of Biological Sciences 15: 683-689.

Schodde, R. 1982. Origin, Adaptation and Evolution of Birds in Arid Australia. In Evolution of the Flora and Fauna of Arid Australia, edited by W.R. Barker and P.J.M. Greenslade, 191-224. Frewville: Peacock Publications.

Serventy, D.L., and A.J. Marshall. 1957. Breeding Periodicity in Western Australian Birds: With an Account of Unseasonal Nestings in 1953 and 1955. Emu 57: 99-126.

Skadhauge, E. 1974. Renal Concentrating Ability in Selected West Australian Birds. Journal of Experimental Biology 61: 269-276.

Slater, P., P. Slater, and R. Slater. 2001. The Slater Field Guide to Australian Birds. Sydney: Reed New Holland.

Smith, M.A. 1987. Pleistocene Occupation in Arid Central Australia. Nature 328: 710-711.

Strahan, R. 1983. The Complete Book of Australian Mammals. Sydney: Angus and Robertson.

_______. 1995. Mammals of Australia. Washington: Smithsonian Institution Press.

Strehlow, T.G.H. 1971. Songs of Central Australia. Sydney: Angus & Robertson.

Taylor, C.R., K. Schmidt-Nielsen and J.L. Raab. 1970. Scaling of Energetic Cost of Running to Body Size in Mammals. American Journal of Physiology 219: 1104-1107.

Thorne, A., R. Grun, G. Mortimer, N.A. Spooner, J.J. Simpson, M. McCulloch, L. Taylor and D. Curnoe. 1999. Australia's Oldest Human Remains: Age of the Lake Mungo 3 Skeleton. Journal of Human Evolution 36: 591-612.

Tindale, N.B. and H.A. Lindsay. 1954. The First Walkabout. London: Longmans, Green and Co.

Twidale, C.R. and H. Wopfner. 1990. Dune Fields. In Natural History of the North East Deserts, edited by M.J. Tyler, C.R. Twidale, M. Davies and C.B. Wells, 45-69. Northfield: Royal Society of South Australia.

Tyler, M.J., A. Edwards and G.R. Johnston. 1990. Amphibians and Reptiles. In Natural History of the North East Deserts, edited by M.J. Tyler, C.R. Twidale, M. Davies and C.B. Wells, 183-188. Northfield: Royal Society of South Australia.

van Beurden, E. 1982. Desert Adaptations of *Cyclorana platycephalus*: an Holistic Approach to Desert-adaptation in Frogs. In Evolution of the Flora and Fauna of Arid Australia, edited by W.R. Barker and P.J.M. Greenslade, 235-239. Frewville: Peacock Publications.

van Oosterzee, P. and R. Morrison. 1991. The Centre: the Natural History of Australia's Desert Regions. Balgowlah: Reed Books.

Watts, C.H.S. 1995. Fawn Hopping-mouse, *Notomys cervinus*. In Mammals of Australia, edited by R. Strahan, 574-575. Washington: Smithsonian Institution Press.

Williams, A.J. and C.R. Dickman. 2004. The Ecology of Insectivorous Bats in the Simpson Desert, Central Australia: Habitat Use. Australian Mammalogy 26: 205-214.

Williams, W.D. 1968. Australian Freshwater Life: the Invertebrates of Australian Inland Waters. Melbourne: Sun Books.

Willoughby, E.J. and M. Peaker. 1979. Birds. In Comparative Physiology of Osmoregulation in Animals, Vol. 2, edited by G.M.O. Maloiy, 1-55. London: Academic Press.

Woolley, P.A. 1984. Reproduction in *Antechinomys laniger* ('*spenceri*' form) (Marsupialia : Dasyuridae): Field and Laboratory Investigations. Australian Wildlife Research 11: 481-489.

_______. 1989. Mulgaras, *Dasycercus cristicauda* (Marsupialia: Dasyuridae); Their Burrows, and Records of Attempts to Collect Live Animals Between 1966 and 1979. Australian Mammalogy 13: 61-64.

Younger, R.M. 1988. Kangaroo: Images Through the Ages. Hawthorne: Century Hutchinson Australia.

INDEX

Printed in the United States
96919LV00004B/4/A

9 781598 583830